INTRODUCTION TO
NANOSCIENCE AND
NANOTECHNOLOGY

WILEY SURVIVAL GUIDES IN ENGINEERING AND SCIENCE

Emmanuel Desurvire, Editor

INTRODUCTION TO NANOSCIENCE AND NANOTECHNOLOGY

Chris Binns

A JOHN WILEY & SONS, INC., PUBLICATION

Published by John Wiley & Sons, Inc., Hoboken, New Jersey
Published simultaneously in Canada

For general information on our other products and services or for technical support, please contact our Customer Care Department within the United States at (800) 762-2974, outside the United States at (317) 572-3993 or fax (317) 572-4002.

Wiley also publishes its books in a variety of electronic formats. Some content that appears in print may not be available in electronic formats. For more information about Wiley products, visit our web site at www.wiley.com.

Library of Congress Cataloging-in-Publication Data:

Binns, Chris, 1954-
 Introduction to nanoscience and nanotechnology / Chris Binns.
  p. cm. – (Wiley survival guides in engineering and science)
 Includes bibliographical references.
 ISBN 978-0-471-77647-5 (cloth)
 1. Nanoscience–Popular works. 2. Nanotechnology–Popular works. I. Title.
  QC176.8.N35B56 2010
  620′.5–dc22
                                                            2009045886

Printed in the United States of America

10 9 8 7 6 5 4 3 2 1

◼◼◼◼ CONTENTS

■■■■■ PREFACE

This book has been a long time in the making. I was taken aback recently when I looked at the original proposal and found that it was written in Spring 2005. At the time there did not appear to be any books that covered the entire field of nanotechnology in a holistic manner written for the layman. To me, one of the most exciting aspects of Nanoscience and Nanotechnology is that they transcend the barriers between the mainstream scientific disciplines of Physics, Chemistry, Biology and Engineering. Thus they provide new insights into the nature of matter and dazzling possibilities for new technology. So full of enthusiasm I put hand to keyboard and embarked on my project to fill this gap, with an original intention to finish in 18 months. Of course the entire sweep of the topic, including, as it does, all the mainstream sciences, is incredibly broad and the intention was to cover it with a relatively light touch to give the reader the 'feel' of what is exciting about the subject. Nanotechnology has a way of sucking you in however and there were so many things that I just couldn't resist including that the touch soon began to get heavier. At some stage in this process of increasing depth I decided to go a stage further and increase the academic level by including 'Advanced Reading Boxes' and some worked problems. This was so that the book could be used as an introductory text for University courses on Nanotechnology, which are becoming increasingly common. Indeed much of the material has been foisted on our own undergraduates at the University of Leicester where we run such a course. The book is still written so that the subject material is covered if one never ventures into these boxes but they provide additional depth for the serious student.

So, four and a half years later, here we are with a longer and more detailed book than I originally intended but in which, I hope, the original intention of giving the reader a holistic feel of the subject has not been lost. There are now many excellent books on Nanotechnology available at a range of different academic levels but I believe that this one still has the widest scope. Despite that, there are still holes, for example, the important areas of Nanomechanics and Nanofluidics, which I hope to fill in future editions. Whatever your use for the book I hope you enjoy it and get from it the excitement that is fundamental to the topic.

CHRIS BINNS
December 2009

ACKNOWLEDGMENTS

I don't think it is possible for a person, at least, a person with a family, to write a book without a good deal of support. In my case I have had massive support from my wife, Angela who has also been a constant source of inspiration and indeed practical help by proof-reading the book. I dedicate this book to her.

I thank my extended family accrued from two marriages, that is, Callum, Rory, Connor, Edward, Tamsyn and Sophie for bringing inspiration and joy to my life as well as the inevitable problems. I would also like to thank my first wife Nerissa for supporting me in my early career and shouldering a large part of the burden of looking after our children.

Finally I would like to thank my parents who with meager resources supported me in pursuit of higher education despite their feeling that I should get a proper job.

1931 Ernst Ruska and Max Knoll build the first Transmission Electron Microscope (TEM) [1]. See chapter 4, section 4.4.6

1959 Lecture by Richard Feynman entitled "There is plenty of room at the bottom, an invitation to enter a new field of physics" [2]. In it he said:

> "...*I am not afraid to consider the final question as to whether, ultimately-in the great future-we can arrange the atoms the way we want; the very atoms, all the way down! What would happen if we could arrange the atoms one by one the way we want them....*" See chapter 4 section 4.4.2

1968 Development of Molecular Beam Epitaxy (MBE) by Arthur and Cho that enables materials to be grown an atomic layer at a time.

1974 N. Taniguchi generally credited for using the word 'Nanotechnology' for the first time [3].

1981 Development of Scanning Tunneling Microscope (STM) by Rohrer and Binnig enables atomic resolution images of surfaces [4]. See chapter 4, section 4.4.1.

1985 Discovery of C_{60} and other fullerenes by Harry Kroto, Richard Smalley and Robert Curl, Jr. [5]. See chapter 3, section 3.2.

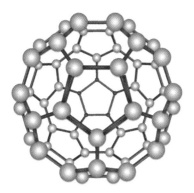

1985 Tom Newman wrote the first page of Charles Dickens' novel *A tale of two cities* with a reduction factor of 25 000 using Electron Beam Lithography (EBL) [6] thus winning a prize of $1,000 offered by Richard Feynman after his 1959 speech. See chapter 4, section 4.2.1.

1986 Development of Atomic Force Microscope (AFM) by Binnig and co-workers [7]. See chapter 4, section 4.4.4.

1987 Development of Magnetic Force Microscope (MFM) by Martin and Wick-ramasinghe [8]. See chapter 4, section 4.4.4.

1990 D. M. Eigler and E. K. Schweizer use an STM to demonstrate atomic-scale positioning of individual Xe atoms on a Ni surface at low temperature (4K) to write "IBM" [9]. This is the first step towards the realization of the Feynman dream set out in the highlighted statement statement from his lecture above. See chapter 4, section 4.4.2.

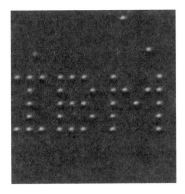

1991 Sumio Iijima discovers carbon nanotubes [10]. See chapter 3, section 3.8

1993 Iijima and Ichihashi grow single-wall carbon nanotubes [11]. See chapter 3, section 3.8.

1995 Takahashi and co-workers demonstrate single-electron transistor operating at room temperature [12]. See chapter 5, section 5.3

1996 Cuberes, Schlittler and Gimzewski demonstrate room temperature positioning of individual C_{60} fullerenes with an STM to produce the "C_{60} abacus" [13]. See chapter 4, section 4.4.2.

1997 Steve Lamoreaux measures the Casimir force at sub-micron distances [14]. See chapter 8, section 8.2

1998 Umar Mohideen and Anushree Roy use AFM used to measure the Casimir force at distance scales down to 90 nm [15]. See chapter 8, section 8.2

2001 Postma and co-workers, demonstrate single-electron transistor operation in a carbon nanotube [16]. See chapter 5, section 5.4.

2002 Regression of tumour in mouse achieved using magnetic nanoparticle hyperthermia achieved by Brusentsov and co-workers [17]. See chapter 6, section 6.2.2.

2007 Johanssen and co-workers conduct first human clinical trials of magnetic nanoparticle hyperthermia treatment of cancer [18]. See chapter 6, section 6.2.2.

REFERENCES

1. See http://www.microscopy.ethz.ch/history.htm.

2. R. P. Feynman, *There is plenty of room at the bottom, an invitation to enter a new field of physics*, Engineering Science Magazine **23** (1960) 143.

3. N. Taniguchi, *On the basic concept of nano-technology*, Proceedings of the International Conference of Production Engineering (Tokyo) Japanese Society of Precision Engineering Part II (1974) 245.

4. G. Binnig and H. Rohrer, *Scanning Tunneling Microscopy*, IBM Journal of Research and Development **30** (1986) 355–369.

5. H. W. Kroto, J. R. Heath, S. C. O'Brien, R. F. Curl and R. E. Smalley, *C60: Buckminsterfullerene Nature* **318** (1985) 162.

6. T. Newman, *Tiny tale gets grand*, Journal of Engineering Science **49** (1986) 24.

7. G. Binnig, C. F. Quate and Ch. Gerber, *Atomic force microscope*, Physical Review Letters **56** (1986) 930–933.

8. Y. Martin and H. K. Wickramasinghe, *Magnetic imaging by "force microscopy" with 1000 Å resolution* Applied Physics Letters **50** (1987) 455.

9. D. M. Eigler and E. K. Shweizer, *Positioning single atoms with a Scanning Tunneling Microscope, Nature* **344** (1990) 524–526.

10. S. Iijima, *Helical microtubules of graphitic carbon*, Nature **354** (1991) 56–58.

11. S. Iijima and T. Ichihashi *Single-shell carbon nanotubes of 1nm diameter*, Nature **363** (1993) 603–605.

12. Y. Takahashi, M. Nagase, H. Namatsu, K. Kurihara, K. Iwdate, Y. Nakajima, S. Horiguchi, K. Murase and M. Tabe, *Fabrication technique for Si single-electron transistor operating at room temperature*, Electronics Letters **31** (1995) 136–137.

13. M. T. Cuberes, R. R. Schlittler and J. K. Gimzewski, *Room-temperature repositioning of individual C_{60} molecules at Cu steps: Operation of a molecular counting device*, Applied Physics Letters **69** (1996) 3016.

14. S. K. Larmoreaux, *Demonstration of the Casimir force in the 0.6 μm to 6 μm range*, Physical Review Letters **78** (1997) 5–8.

15. Umar Mohideen and Anushree Roy, *Precision measurement of the Casimir force from 0.1 to 0.9μm*, Physical Review Letters **21** (1998) 4549.

16. H. W. Ch. Postma, T. Teepen, Z. Yao, M. Grifoni and C. Dekker, *Carbon nanotube single-electron transistors at room temperature*. Science **293** (2001) 76–79.

17. N. A. Brusentsov, L. V. Nikitin, T. N. Brusentsova, A. A. Kuznetsov, F. S. Bayburtskiy, L. I. Shumakov and N. Y. Jurchenko, Journal of Magnetism and Magnetic Materials **252** (2002) 378–380.

18. M. Johanssen, U. Gneveckow, K. Taymoorian, B. Thiesen, N. Waldöfner, R. Scholz, K. Jung, A. Jordan, P. Wust and S. A. Loening, *Morbidity and quality of life during thermotherapy using magnetic nanoparticles in locally recurrent prostate cancer: Results of a prospective phase I trial''*, International Journal of Hyperthermia, **23** (2007) 315–323.

INTRODUCTION

Research in nanotechnology is a growth industry, with worldwide government-funded research spending running at over four billion dollars per year and growing at an annual rate of about 20% [1]. Industry is also willing to spend vast sums on investigating nanotechnology, with, for example, major cosmetics companies announcing big increases in their annual Research and Development budgets for the field. It is clear that nanotechnology is expected to have a significant impact on our lives, so what is it and what does it do? These simple direct questions, unfortunately, do not have simple direct answers, and it very much depends on who you ask. There are thousands of researchers in nanotechnology in the world, and one suspects that one would get thousands of different responses. A definition that would probably offend the smallest number of researchers is that nanotechnology is the study and the manipulation of matter at length scales of the order of a few nanometers (100 atoms or so) to produce useful materials and devices.

This still leaves a lot of room for maneuver. A nanotechnologist working on suspensions of particles might tell you that it is achieving better control of tiny particles a few nanometers across (nanoparticles) so that face creams can penetrate the epidermis (outer skin layer). A scientist working at the so-called "life sciences interface" would say that it is finding ways of attaching antibodies to magnetic nanoparticles to develop revolutionary cancer treatments. A researcher working on "molecular electronics" will tell you that it is creating self-ordered assemblies of nanoparticles to produce electronic circuits in which the active components are a thousand times smaller than a single transistor on a Pentium IV chip. Some nanotechnologists (a small minority) would tell you that it is finding ways to build tiny robots whose components are the size of molecules (nanobots).

We will talk in detail about size scales in Chapter 1; but for the moment consider Fig. 0.1, which shows, schematically, the size scale of interest in nanotechnology (the nanoworld). For reasons that will become clear in Chapter 1, the upper edge of the nanoworld is set at about 100 nm. Even though this is hundreds of times smaller than the tiniest mote you can see with your eyes and is smaller than anything that can be resolved by the most powerful optical microscope, a chunk of matter this size or bigger can be considered to be a "chip off the old block"—that is, a very tiny piece of ordinary material. If we were to assemble

Introduction to Nanoscience and Nanotechnology, by Chris Binns
Copyright © 2010 John Wiley & Sons, Inc.

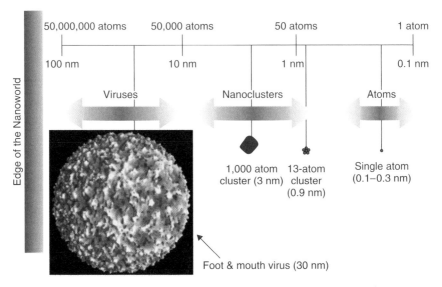

Fig. 0.1 The nanoworld. The size range of interest in nanotechnology and some representative objects.

pieces of copper or iron this big into a large chunk, the resulting block would behave exactly as we would expect for the bulk material. Thus nanotechnology does not consider pieces of matter larger than about 100 nm to be useful building blocks.

As shown in Fig. 0.1, viruses are small enough to be inhabitants of the nanoworld whereas bacteria are much larger, being typically over 10 μm (10,000 nm) in size, though they are packed with "machinery" that falls into the size range of the nanoworld (see Chapter 6, Section 6.1.3). Going down in size, the figure shows typical sizes of metal particles, containing ~1000 atoms that can be used to produce advanced materials. The properties of these (per atom) deviate significantly from the bulk material, and so assembling these into macroscopic chunks produces materials with novel behavior.

Finally, the lower edge of the nanoworld is defined by the size of single atoms, whose diameters vary from 0.1 nm (hydrogen atom) to about 0.4 nm (uranium atom). We cannot build materials or devices with building blocks smaller than atoms, and so these represent the smallest structures that can be used in nanotechnology.

There are so many aspects to nanotechnology that one of the difficulties in writing about it is finding ways to organize the text into a coherent structure. This book will largely follow a classification scheme introduced by Richard Jones in his book *Soft Machines Nanotechnology and Life* [2] that helps to categorise nanotechnology into a logical framework. He defines three categories in order of increasing sophistication—that is, *incremental, evolutionary*, and *radical* nanotechnology. These are described in detail below.

0.1 INCREMENTAL NANOTECHNOLOGY

All substances, even solid chunks of metal, have a grain structure, and controlling this grain structure allows one to produce higher performance materials. This could mean stronger metals, magnetic films with a very high magnetization, suspensions of nanoparticles with tailored properties, and so on. Actually, some aspects of incremental nanotechnology can be considered to date back to the ancients. For example, the invention of Indian ink, probably in China around 2700 B.C., relies on producing carbon nanoparticles in water. Also medieval potters in Europe knew how to produce a lustre on pots by coating them with copper and silver nanoparticles [3], a process that can be traced back to 9th century A.D. Mesopotamia. Figure 0.2 shows an electron microscope image of the glaze of a 16th-century Italian pot, whose luster derives from the coating by 5-nm-diameter copper particles.

Most modern nanotechnologists would be proud of the size control of the particles in this picture. Whereas these days a process that involved nanoparticles such as this would be proudly claimed to be nanotechnology and thus open the door

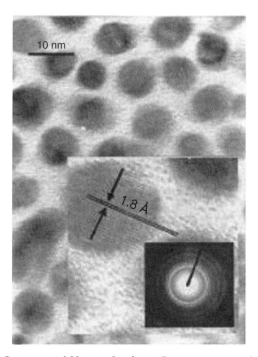

Fig. 0.2   Ancient Incremental Nanotechnology. Copper nanocrystals on a 10th century pot of about 10 nm diameter used to produce a surface luster. The inset shows an increased magnification image of a single 7 nm diameter particle with atomic planes visible revealing its crystallinity. Reproduced with the permission of Elsevier Science from I. Borgia et al. [3].

to research funding, spin-off companies, and so on, the ancients were developing processes that did something invisible to the materials but nevertheless allowed them to achieve certain results. In this sense, a lot of incremental nanotechnology can sometimes be considered to be a re-branding of other, more traditional lines of research such as materials science and chemistry. The nanotechnology title is still useful, however, since nanotechnology is, by its nature, multidisciplinary and it encourages cross-disciplinary communication between researchers.

The aspect of incremental nanotechnology that has really changed in the modern world is the development of instruments (see Chapter 4) that can probe at the nanoscale and image the particles within materials or devices. Researchers can actually observe what is happening to the particles or grains in response to changes in processing. This not only makes development of new processes more efficient but also leads to the discovery of completely new structures that were not known to exist and hence new applications. Nature is full of surprises when one studies sufficiently small pieces of matter, as will become clear throughout this book.

0.2 EVOLUTIONARY NANOTECHNOLOGY

Whereas *incremental nanotechnology* is the business of assembling vast numbers of very tiny particles to produce novel substances, *evolutionary nanotechnology* attempts to build nanoparticles that *individually* perform some kind of useful function. They may need to be assembled in vast numbers to form a macroscopic array in order to produce a device, but a functionality is built into each one. Such nanoparticles are necessarily more complex than those used in incremental nanotechnology. A simple example is a magnetic nanoparticle that is used to store a single bit of information by defining the direction of its magnetization. If one wants to do this using nanoparticles with diameters smaller than about 6 nm at ambient temperature, simple elemental nanoparticles made, for example, of pure Fe will not work because thermal vibrations will instantly change their direction of magnetization. To produce a particle that doesn't lose its "memory" without cooling to very low temperatures, each one has to be made with more than one material and formed either as a uniform alloy or as a core-shell particle (see Fig. 0.3a)—that is, a kind of nanoscale chocolate peanut with a core consisting of one material surrounded by a shell of a different substance. As stated above, these have to be assembled in vast numbers into some sort of ordered array (the big unsolved problem with this technology) to produce a useful device, but each particle has within it the capacity to store a data bit. If and when this technology succeeds, it would represent a storage density about 1000 times greater than existing hard disks on computers. For example, the bottom image in Fig. 0.3a shows an array of core-shell nanoparticles used to store the word "nanotechnology" in ASCII code. The storage density represented would allow about two million books or a large library to be written in an area the size of a postage stamp.

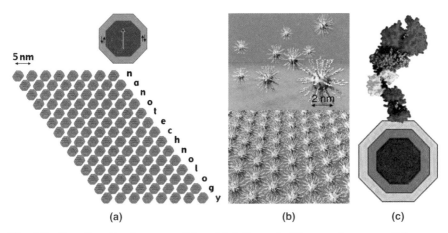

(a)   (b)   (c)

Fig. 0.3   Functionalized nanoparticles. (a) A "core-shell" magnetic nanoparticles consisting of a ferromagnetic metal core (e.g. iron or cobalt) surrounded by an antiferromagnetic shell, e.g. cobalt-oxide or manganese) that is specialized to record a single bit of
information encoded by the direction of the core magnetization (see Chapter 5, Section
5.1). The lower picture shows an array storing the word 'nanotechnology' in ASCII format
with blue representing down magnetization or '0' and red representing up magnetization
or '1'. Writing at this density would enable the storage of about two million books (a
large University library) on an area the size of a postage stamp. (b) Gold nanoparticles
with attached thiol molecules. Each nanoparticle can behave as a transistor and the thiols
can bond onto other thiol-coated gold nanoparticles via electrically resistive or capacitive
links to build circuits with a component density 1000 times greater than existing devices
(see Chapter 5, Section 5.3). Reproduced with the permission of Dr. Mark Everard from
[4]. (c) A core-shell magnetic nanoparticle (as in (a)) with a second shell of gold that
makes it easy to attach biological molecules such as proteins or antibodies, or drugs. The
magnetic core of the particle can be utilized to steer the attached molecule to specific
areas of the body by external fields for targeted drug delivery. Alternatively, the attached
biological molecule could be used to target specific cells (e.g. tumor cells) that could then
be heated and killed by a weak external radio-frequency magnetic field that is harmless
to healthy tissue (see Chapter 6, Section 6.2.2).

An example of a more sophisticated "functionalized" nanoparticle, shown in
Fig. 0.3b, is a gold nanoparticle about 2 nm across with attached molecules
called *thiols*. If wires could be attached to this nanoparticle, it could be made
to act like a transistor by a process called Coulomb blockade (see Chapter 5,
Section 5.3). It turns out that the thiols, of which there are many types, can
act as wires if they come together in the right way. A slight change in the
bonding produces a change in the resistance of the link or makes it capacitive
(insulating). In other words, an entire circuit network consisting of transistors,
capacitors, and resistors can be produced by placing an array of thiol-coated
nanoparticles in the correct positions. This, of course, is the unsolved problem
but is a tantalizing one because the density of components in such an array means

that about 1000 nanoparticle transistors could be placed in the space occupied by a single silicon-based transistor on a Pentium IV chip.

No one can fail to be impressed by the huge increases in performance and density of components/memory elements in devices made by the electronics and magnetic recording industries in the last few decades. The above two examples illustrate, however, that there is still a long way to go, nicely reinforcing a lecture on nanotechnology given by the visionary Nobel laureate Richard Feynman and entitled *There's Plenty of Room at the Bottom*. The amazing thing about this lecture was that it was given in 1959.

Continuing the trend toward complexity of individual nanoparticles, Fig. 0.3c shows a combination of the types in Figs. 0.3a and 0.3b consisting of a magnetic core-shell particle, with controlled properties, coated with a second shell of gold that facilitates its attachment to complex biological molecules—for example, drugs, proteins, or antibodies. The magnetic core of the particle can be utilized to steer the attached molecule to specific areas of the body by external fields for targeted drug delivery. Alternatively, the attached biological molecule could be used to target specific cells (e.g., tumor cells) that could then be heated and killed by a weak external radio-frequency field that is harmless to healthy tissue (see Chapter 6, Section 6.2.2).

As a rough guide to where we are with evolutionary nanotechnology, the functional nanoparticles shown in Fig. 0.3 can be routinely manufactured, but their use in technologies such as those described above awaits the solution to enormously difficult technological problems such as controlling their self-assembly into arrays.

0.3   RADICAL NANOTECHNOLOGY

Finally, the most far-reaching version of nanotechnology, described as *radical nanotechnology* by Richard Jones, is the construction of machines whose mechanical components are the size of molecules. The field has bifurcated into two distinct branches, that of (a) *molecular manufacturing* in which macroscopic structures and devices are built by assembling their constituent atoms, and (b) nanorobots or *nanobots*, which are invisibly small mobile machines. Molecular manufacturing was originally proposed by the Nobel laureate Richard Feynman in his famous lecture in 1959 and was subsequently advocated with much enthusiasm by Eric Drexler [5]. In 1990 the IBM research laboratories in Zurich demonstrated that they could move and position individual atoms using a scanning tunneling microscope (STM—see Chapter 4, Section 4.4.2), lending support to the idea that molecular manufacturing may, at least in principle, be possible. The problems and the emergence of some enabling technologies for molecular manufacturing are presented in Chapter 7.

Nanobots have generated a good deal of controversy, especially ones that can play atomic Lego and build anything out of atoms lying around. If this were possible, then one could, in principle, build a nanobot that moved around exploring the surface it occupied. If it were equipped with an assembler that could assemble atoms and molecules, it could make a copy of itself by rooting around and finding the atoms it needed to reproduce. Obviously, this kind of activity would need on-board intelligence, and this could be provided by either a mechanical computer, again with molecular-sized components, or something more akin to a the molecular electronic-type circuit shown in Fig. 0.3b. Since each nanobot could make multiple copies of itself, the population could increase exponentially and would quickly produce a sufficiently vast army to build macroscopic objects. Drexler himself pointed out the doomsday scenario where the nanobots multiply out of control like a virus and eventually exist in such vast numbers that they could rearrange the atoms of the planet to produce a kind of "gray goo." Unfortunately, this scenario has tended to hijack discussions on radical nanotechnology; and since the two branches of radical nanotechnology have been melded together in the public debate, there is a general feeling that all radical nanotechnology is dangerous. The reality is that exponentially self-replicating machines are not required for molecular manufacturing [6] and nanobots do not need to be built with assemblers to self-replicate in order to perform useful functions.

There is a scientific debate about whether this technology is feasible, even in the long term, or indeed desirable, but the discussion has moved on from generalities to a consideration of the detailed processes required for molecular manufacturing (see Chapter 7 and the references therein). A frequently proposed argument in favor of radical nanotechnology is that it already exists in all living things. Biological cells are filled with what may be regarded as nanomachines and molecular assemblers. Biology, however, is very different to the nanoscale process-engineering path envisaged by radical nanotechnologists, as explained in *Soft Machines Nanotechnology and Life* [2]. It is fair to say that both the feasibility and timescale of Radical Nanotechnology divides the community. The point is that while incremental nanotechnology exists and evolutionary nanotechnology is close (~10 years), radical nanotechnology, if feasible, is probably decades away. Whatever the twists and turns of the debate, once we get away from the argument over nanobots, there is no doubt that the ability to produce nanomachines and achieve safe nonexponential molecular manufacturing will reap enormous benefits.

It is possible that the solution to some of the more difficult technological problems involved with radical nanotechnology may arise from a better fundamental understanding of the true nature of empty space. The quantum description of our universe predicts new types of force at very short distance scales (nanometers) arising directly out of the properties of vacuum. Although we can only detect these forces with very sensitive instruments (the tools of nanotechnology in fact—see Chapters 4 and 8), to a nanoscale machine whose components are within nanometers of each other, these forces will be as natural a part of their environment as gravity is to us. Research on these forces and how to utilize them in nanotechnology is already being undertaken by several research groups worldwide. This may be the missing link between biology and radical nanotechnology—that is, natural systems, whose inner workings happen on the same scale as nanomachines have evolved over billions of years and must have utilized all available forces including the exotic ones.

0.4   BOTTOM-UP/TOP-DOWN NANOTECHNOLOGY

Finally in this introduction it is worth mentioning another way of categorizing nanotechnology, that is, *bottom-up* and *top-down* approaches. Everything discussed so far has been part of a bottom-up approach in which the building block (nanoparticle, molecular machine component, etc.) is identified and produced naturally and then assembled to produce the material or device required. In the top-down approach, you start with a block of some material and machine a device or structure out of it. This is akin to conventional engineering using lathes and millers to machine a shape out of a solid block. The modern tools of nanotechnology, however, are able to machine structures with sizes of a few nanometers, so the size of components made with a top-down approach is not much different from the building blocks of the bottom-up approach. The flexibility of top-down tools—in particular, focused ion beam systems (FIBs)—is further enhanced by their ability to deposit material to produce nanoscale features as well as to remove it. This is beautifully illustrated in Fig. 0.4, which shows an example of a "wineglass" with a cup diameter 20 times smaller than the width of a human hair produced by depositing carbon. Although this is a rather big structure on the scale of nanometers, the smallest feature size that can be produced by a modern FIB is less than 100 nm.

The two approaches (top-down and bottom-up) are complementary, and some of the most exciting research arises out of combining them. For example, if one wants to measure the electrical or magnetic properties of an individual nanoparticle, the fantastic precision of a modern top-down tool enables the production of electrodes that can attach to it. In general terms, the bottom-up/top-down categorization can be applied separately to each of the incremental, evolutionary, and radical nanotechnology categories.

The above is an attempt at a lightning tour of nanotechnology with generic descriptions and without addressing details. The rest of the book looks in detail at these and other aspects of nanotechnology. Chapter 1 aims to instill a feeling of how small the nanometer length scale is in comparison to macroscopic objects and why it is special. It discusses the basic conception of the discrete nature of matter starting from the original philosophical ideas of Leucippus and Democritus of ancient Greece to the modern view of atomic structure. It also describes why the properties of pieces of matter with a size in the nanometer range (nanoparticles) deviate significantly from the bulk material and how these special properties may be used to produce high-performance materials and devices. In Chapter 2 the discussion is broadened to include naturally occurring nanoparticles, both in the Earth's atmosphere and in space. Chapter 3 is dedicated to nanoparticles composed of carbon, and the justification for devoting a chapter to a single element is the rich variety of nanostructures produced by carbon and

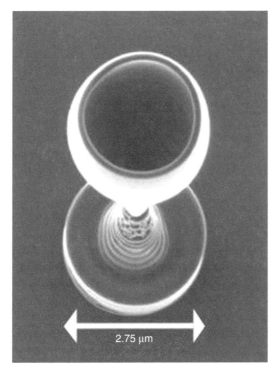

Fig. 0.4   The smallest wineglass in the world (authorized by *Guinness World Records*). Wine glass whose cup diameter is 20 times smaller than the width of a human hair produced by deposition of carbon using a focused ion beam (FIB) machine. The structure arose from a Joint development by SII NanoTechnology, NEC, and the University of Hyogo, Japan. Although this is a rather big structure on the scale of nanometers, the smallest feature size that can be produced by a modern FIB is less than 100 nanometers (see Chapter 4, Section 4.2.2).

their importance in the rest of nanotechnology. Chapter 4 presents the tools of nanotechnology that can build, image, and manipulate nanostructures to build materials and devices using a bottom-up approach. It also describes top-down manufacturing methods that are capable of shaping nanostructures and perhaps one of the most exciting aspects of the field—that is, combining bottom-up and top-down approaches so that *individual* nanostructures can be probed. Chapter 5 is about artificially produced nanostructures that have a built-in functionality. Examples presented include magnetic nanoparticles that can store a data bit, nanoparticles that can function as transistors and quantum dots, which behave as "artificial atoms" with novel optical and electronic properties. Chapter 6 shows how combining advances in the production of nanoparticles and in biotechnology makes it possible to produce biologically active nanoparticles that can interact with specific cells in the body. These can then be used as nanoscale-amplifiers in biological images or they can be used to destroy their attached cells under the application of external stimulation, such as microwave or infrared radiation leading to powerful new treatments for cancer. Chapter 7 presents radical nanotechnology and discusses the potential for building autonomous machines with nanoscale components. Chapter 8 discusses how the tools of nanotechnology can be exploited to study the basic nature of vacuum itself via the Casimir effect. This is a strange "force from nothing" that arises from the zero-point energy density of empty space. These experiments may one day uncover a deeper level to our universe that underlies the observable universe consisting of all particles and all normal energy that we can sense or detect with instruments. The Casimir force may also be an important phenomenon for the practical implementation of nanomachines as a method to transmit force without contact.

The three-tier classification scheme divides among the chapters as follows. Chapters 1 and 3 deal with incremental nanotechnology, Chapters 5 and 6 deal with evolutionary nanotechnology, and Chapter 7 deals with radical nanotechnology. Where appropriate, a separate bottom-up/top-down categorization is introduced. Chapter 2 deals with naturally occurring nanoparticles, Chapter 4 deals with the tools of nanotechnology, and Chapter 8 deals with the Casimir force, so these stand outside the broad classification scheme.

REFERENCES

1. *Nanoscience and Nanotechnologies: Opportunities and Uncertainties*, Report by the Royal Society and Royal Academy of Engineering, published July 29, 2004. Available from http://www.nanotec.org.uk/finalReport.htm.
2. R. A. L. Jones, *Soft Machines Nanotechnology and Life*, Oxford University Press, Oxford, 2004.
3. I. Borgia, B. Brunetti, I. Mariani, A. Sgamelotti, F. Cariati, P. Fermo, M. Mellini, C. Viti, and G. Padeletti, Heterogeneous distribution of metal nanocrystals in glazes of historial pottery, *Appl. Surface Sci.* **185** (2002), 206–216.
4. M. J. Everard, X-ray scattering studies of self-assembled nanostructures, PhD thesis, 2003, University of Leicester.
5. E. Drexler, *Engines of Creation*, Garden City, NY, 1986 and Fourth Estate, London, 1990.
6. C. Pheonix and E. Drexler, Safe exponential manufacturing, *Nanotechnology* **15** (2004), 869–872.

CHAPTER 1

Size Matters

1.1 THE FUNDAMENTAL IMPORTANCE OF SIZE

The aim of this chapter is to instill an intuitive feel for the smallness of the structures that are used in nanotechnology and what is special about the size range involved. As we will see, the importance of the nanoscale is wrapped up with fundamental questions about the nature of matter and space that were first pondered by the ancient Greeks. Three thousand years ago, they led the philosophers Leucippus and Democritus to propose the concept of the atom. These ideas will come around full circle at the end of the book when we will see that modern answers (or at least partial answers—the issue is still hot) are very much wrapped up in nanotechnology.

First of all, let us remind ourselves how small nanostructures really are. This is a useful exercise even for professionals working in the field. The standard unit of length in the metric system is the meter, originally calibrated from a platinum–iridium alloy bar kept in Paris but since 1983 has been defined as the distance that light travels in 1/299,792,458 seconds. For convenience, so that we are not constantly writing very small or very large numbers, the metric system introduces a new prefix every time we multiply or divide the standard units by 1000. Thus a thousandth of a meter is a *milli*-meter or mm (from the Roman word *mille* meaning 1000); a thousandth of a millimeter (or a millionth of a meter) is a *micro*-meter or μm (from the Greek word *mikros* meaning small). Similarly, a thousandth of a micrometer (or a billionth of a meter) is a *nano*meter or nm (from the Greek word *nanos* meaning dwarf). As shown in Fig. 0.1 ("the nanoworld"), we will be dealing in building blocks that vary from 100 nm across down to atoms, which are 0.1 to 0.4 nm across.

Nowadays, instruments can directly image nanostructures; for example, Fig. 1.1 shows a scanning tunneling microscope (STM—see Chapter 4, Section 4.4.1) image of a few manganese nanoparticles with a diameter of about 3 nm

Introduction to Nanoscience and Nanotechnology, by Chris Binns
Copyright © 2010 John Wiley & Sons, Inc.

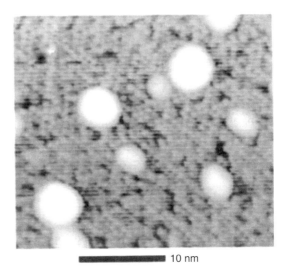

10 nm

Fig. 1.1   Manganese nanoparticles on bucky balls. STM image (see Chapter 4, Section 4.4.1) of a few manganese nanoparticles with a diameter of about 3 nm (i.e. each one contains a few hundred manganese atoms) deposited onto a bed of carbon-60 nanoparticles with a diameter of 0.7 nm on a silicon surface. Reproduced with the permission of the American Institute of Physics from M. D. Upward et al. [1].

deposited onto a bed of carbon-60 molecules ("bucky balls"—see Chapter 3) with a diameter of 0.7 nm on silicon. It thus displays two different nanoparticles of interest in the same image. This could just as easily be a picture of snowballs on a bed of marbles, and it is easy to lose sight of the difference in scale between our world and that of the nanoparticles. To get some idea of the scale, take a sharp pencil and gently tap it onto a piece of paper with just enough force to get a mark that is barely visible. This will typically be about 100 μm or 100,000 nm across. If the frame in Fig. 1.1 were this large, it would contain about 100,000,000 of the Mn nanoparticles and 20,000,000,000 of the carbon-60 nanoparticles, which is about three times the population of the planet.

In fact, distance scales used in science go to much smaller than nanometers and much larger than meters, but the title of this book suggests that there is something special about the nanoeter scale—so what is it? The answer to this lies in philosophical questions that were originally posed regarding the fundamental nature of matter and space by Democritus and his contemporaries that still resonate today. Democritus (Fig. 1.2) was born in Abdera, Northern Greece in about 460 B.C. to a wealthy noble family, and he spent his considerable inheritance (millions of dollars in today's currency) traveling to every corner of the globe learning everything he could. Known as the laughing philosopher, he lived to over 100 years old, so it appears to have been a good life. He wrote more than 75 books about almost everything from magnets to spiders and their webs. Only fragments of his work survived, with most of his books being destroyed in the

Fig. 1.2 Democritus. Marble bust of Democritus, Victoria and Albert Museum, London. Reproduced with permission from Carlos Parada, Greek Mythology Link: (http://homepage.mac.com/cparada/GML).

third and fifth centuries. His lasting impact on modern science was to propose, with his teacher Leucippus, the concept of the atom.

All our experience in the macroscopic world suggests that matter is continuous; and thus with nothing but our eyes for sensors, the original suggestion that matter is made from continuous basic elements such as earth, fire, air, and water seems reasonable. This, however, leads to a paradox because if matter were a continuum, it could be cut into smaller and smaller pieces without end. If one were able to keep cutting a piece of matter in two, each of those pieces into two, and so on ad infinitum, one could, at least in principle, cut it out of existence into pieces of nothing that could not be reassembled. This led Leucippus and Democritus to propose that there must be a smallest indivisible piece of substance, the *a-tomon* (i.e., uncuttable), from which the modern word *atom* is derived. They suggested that the different substances in the world were composed of atoms of different shapes and sizes, which is not an unrecognisable description of modern chemistry. Once you propose atoms, however, you automatically require a "void" in which they move; and the void is a concept that also produces dilemmas, which were the subject of much debate three thousand years ago. For example, is the void a "something" or a "nothing" and is it a continuum or does it also have a smallest uncuttable piece? Whereas atoms are a familiar part of the scientific world, the true nature of the void between them is something that is still not fully understood and many scientists believe that the void question lies at the heart of all the "big" questions about the universe and the nature of reality. It is a question that nanotechnology can address, and we will return to this discussion in Chapter 8.

Meanwhile returning to atoms, one could argue that the basic philosophy outlined above, which led to their being proposed, means that you should really attribute the *a-tomon* to more fundamental constituents of atoms, such as electrons and quarks. There is a good reason to stick with atoms, however, since we are talking about constituents of materials; and if we pick on a particular material, say copper, the smallest indivisible unit of "copperness" is the copper atom. If we divide a copper atom in two, we get two atoms of different materials.

So what has all this got to do with nanotechnology? Nowadays we can carry out, in practice, the Democritus mind experiment and study pieces of matter of smaller and smaller size right down to the atom. The important result is that the properties of the pieces start to change at sizes much bigger than a single atom. When the size of the material crosses into the "nanoworld" (Fig. 0.1), its fundamental properties start to change and become dependent on the size of the piece. This is, in itself, a strange thing because we take it for granted that, for example, copper will behave like copper whether the piece is a meter across or a centimeter across. This is not the case in the nanoworld, and the onset of this strange behavior first shows up at the large end of the nanoworld scale with the magnetic properties of metals such as iron. It is worth spending a little time on this because it is a clear illustration of how the behavior of a piece of material can become critically dependent on its size.

1.2   THE MAGNETIC BEHAVIOR OF NANOPARTICLES

It is widely known that iron is a magnetic material, but in fact a piece of pure (or "soft") iron is not magnetized. This is easy to prove by taking a piece of soft Fe and seeing that it does not attract a ball bearing (Fig. 1.3a). In contrast, a permanent magnet, which is an alloy, such as neodymium–iron–boron that is permanently magnetized, strongly attracts the ball bearing (Fig. 1.3b). A simple and illustrative experiment is to sandwich the ball bearing between the permanent magnet and piece of soft iron and then pull the magnet and the pure iron apart (Fig. 1.3c). Oddly, while the ball bearing shows no attraction to the soft iron on its own, in the presence of the magnet it stays glued firmly to the piece of soft iron as it is pulled away, showing that it is magnetized to a greater degree than the actual magnet. Beyond a certain distance from the magnet, the soft iron reverts to its demagnetized state and the ball bearing comes loose (Fig. 1.3d).

The source of magnetism in materials is their constituent atoms, which consist of tiny permanent dipolar magnets whose strength is given by the *magnetic moment*[1] of the atom (see Advanced Reading Box 1.1). In a material such as iron, there is a strong interaction (the exchange interaction) between the atoms

---

[1]Magnetic moment or magnetic dipole moment, denoted by the symbol μ, summarizes the characteristics of a simple loop of current, producing a magnetic field, by the equation $\mu = IA$, where I is the circulating current and A is the area enclosed by the loop. A simple loop such as this produces a *dipolar* magnetic field similar to that produced by a bar magnet with *north* and *south* poles.

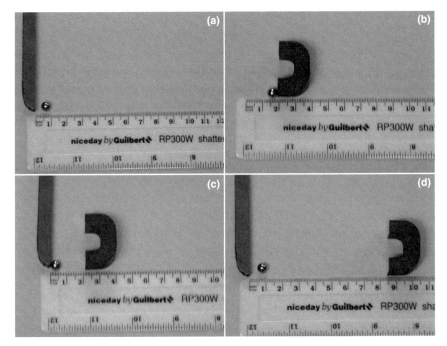

Fig. 1.3 Simple experiment to demonstrate magnetic domains. (a) Soft iron does not attract the ball bearing. (b) A conventional magnet does, however (c) when the soft iron is magnetized by being in the presence of the magnet it becomes more magnetic than the magnet and the ball bearing stays with the soft iron in preference to the magnet. (d) The situation persists until the magnet is far enough away that the soft iron reverts to its domain structure and externally generates no magnetic field.

that lines up the atomic magnets to produce a macroscopic magnetization. Note that the exchange interaction is a quantum mechanical effect and is not the normal interaction that you would see between two bar magnets, for example. For one thing the interaction between bar magnets aligns them in opposite directions, and for another the exchange interaction is thousands of times stronger than the direct magnetic interaction.

ADVANCED READING BOX 1.1—ATOMIC MAGNETIC MOMENTS AND THE EXCHANGE INTERACTION

The individual atoms of most elements have a permanent magnetic moment, so they generate a dipolar magnetic field similar to a simple bar magnet. The source of the atomic magnetic moment is twofold. It arises from the orbital motion of the electrons around the nucleus, which can be considered to constitute a simple current loop, and also from the intrinsic angular momentum (spin) of the electrons. These two contributions generate an orbital and a spin magnetic moment; and for the elements Fe, Co, and Ni, the two contributions are simply added to obtain the total magnetic moment. The exchange interaction that acts between neighboring atoms arises from the Pauli Exclusion Principle. This tends to keep electrons apart if they have the same spins so that the Coulomb repulsion energy between the outermost electrons of neighboring atoms is reduced if the electrons align their spins in the same direction. This appears as a very strong magnetic interaction trying to align the spin magnetic moments, but it is an electrostatic effect produced by the quantum nature of the electrons. It is typically 3–4 orders of magnitude stronger that the direct magnetic interaction of the atomic magnetic moments taken to be simple bar magnets.

So in a magnetic material the powerful exchange interaction tries to line up all the microscopic atomic magnets to lie in the same direction. This, however, is not necessarily the preferred configuration because the uniformly magnetized state generates a magnetic field that passes through the material and the magnetization finds itself pointing the wrong way in its own magnetic field; that is, it has the maximum *magnetostatic* energy.[2] Of course, reversing the magnetization is of no use because the generated field reverses and again the sample magnetization and the generated field are aligned in the least favorable direction to minimize energy. The exchange interaction and the magnetostatic energy are thus competing, which at first glance does not appear to be much of a competition considering that the exchange energy per atom between nearest neighbors is 3–4 orders of magnitude stronger than magnetostatic one. The magnetostatic interaction, however, is long range while the exchange interaction only operates between atomic neighbors. There is thus a compromise that will minimize the energy relative to the totally magnetized state by organizing the magnetization into so-called *domains* with opposite alignment (Fig. 1.4a). If these domains have the right size, the reduction in magnetostatic energy is greater than the increased exchange energy from the atoms along the boundaries that are neighbors and have their magnetization pointing in opposite directions. In the minimum energy state the material does whatever is necessary to produce no external magnetic field, and this is what has happened in Fig. 1.3a. The magnetization of the soft iron has organized itself into domains, and externally it is as magnetically dead as a piece of copper. The actual magnet has been treated to prevent the domains forming so that it stays magnetized (Fig. 1.3b). When we bring the piece of soft iron into the field of the magnet, its domains are all aligned in the same direction and it has a greater magnetization than the magnet so that when we pull the two apart, the ball bearing stays stuck firmly to the soft iron. This continues until the soft iron is far enough away from the magnet to revert to its domain structure and become magnetically dead externally.

---

[2]Magnetostatic energy is the energy of a permanent magnet interacting with a static magnetic field.

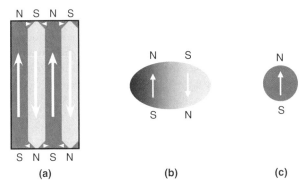

Fig. 1.4   Single-domain particles. Domain formation in iron to minimize energy. Below a critical size (approx. 100 nm), the energy balance favors just a single domain and the piece of iron stays permanently and fully magnetized.

The phrase "*If these domains have the right size*" in the previous paragraph encapsulates the critical point. If we do the Democritus experiment and start chopping the piece of soft iron into smaller and smaller pieces, the number of domains within the material decreases (Fig. 1.4b). There must come a size, below which the energy balance that forms domains simply does not work any more and the particle maintains a uniform magnetization in which all the atomic magnets are pointing the same way (Fig. 1.4c). So what size is this? It turns out to be about 100 nm—that is, the upper edge of the nanoworld. Any iron particle that is smaller than this is a single domain and is fully magnetized. This may seem like a subtle size effect, but it has profound consequences. Fully magnetized iron is a much more powerful magnet than any actual magnet as shown in Fig. 1.3. The reason is that a permanent magnet must contain some nonmagnetic material to prevent the process of domain formation so that its magnetization is diluted compared to the pure material. A world in which every piece of iron or steel was fully magnetized would be very different from our familiar one. Every steel object would attract or repel every other one with enormous force. Cars with their magnetization in opposite directions would be very difficult to separate if they came into contact.

Nature makes good use of this magnetic size effect. Bacteria, such as the one shown in Fig. 1.5, have evolved, which use strings of magnetic nanoparticles to orient their body along the local magnetic field lines of the Earth. The strain shown in the figure, which is found in northern Germany, lives in water and feeds off sediments at the bottom. For a tiny floating life form such as this, knowing up and down is not trivial. If the local field lines have a large angle to the horizontal, as they do in Northern Europe, then the string of magnetic nanoparticles makes the body point downwards and all the bacterium has to do is to swim knowing that it will eventually find the bottom.

The intelligence of evolution is highlighted here. If the particles are single-domain particles, then they will stay magnetized forever, so forming a string of

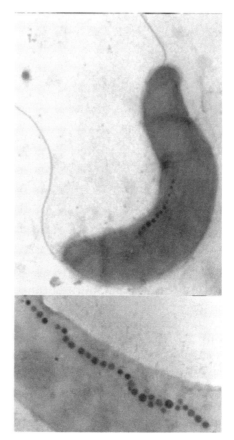

Fig. 1.5 Magnetic bacterium using single-domain particles. The Magnetic bacterium (*magnetospirillum gryphiswaldense*) from river sediments in Northern Germany. The lines of (permanently magnetized) single-domain magnetic nanoparticles, appearing as dark dots, align the body of the bacterium along the local direction of the Earth's magnetic field, which in Germany is inclined at 55° from horizontal. This means that the bacterium will always swim downwards towards the sediments where it feeds. Reproduced with With kind permission of Springer Science and Business Media from D. Schüler [2].

these ensures that the navigation system will naturally work. If the bacterium formed a single piece of the material the same size as the chain of particles, then a domain structure would form and it would become magnetically dead. The nanoparticles are composed of magnetite (Fe_3O_4) rather than pure iron, but the argument is the same. There is currently research devoted to persuading the bacteria to modify the composition of the nanoparticles by feeding them with cobalt-containing minerals as a method of high-quality nanoparticle synthesis (see Chapter 4, Section 4.1.8).

Interestingly, chains of magnetic nanoparticles with a similar structure have been found on a piece of meteorite known to have come from Mars [3]. Since

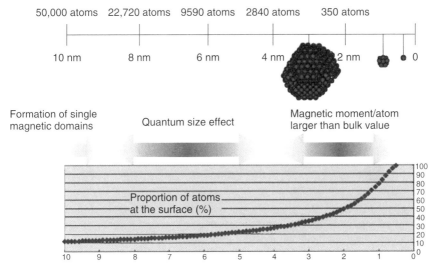

Fig. 1.6   Size-dependent behavior in nanoparticles. For particles smaller than 10 nm, quantum effects start to become apparent. In this size range the proportion of atoms that constitute the surface layer starts to become significant reaching 50% in 2 nm diameter particles. Below about 3 nm the strength of magnetism per atom starts to increase.

the only known way of producing this mineral is biological, the observation is evidence that there was once life on Mars, though this analysis remains controversial. In fact, Mars no longer has a significant planetary magnetic field, which disappeared in the distant past, but supporters of the Martian bacteria proposal argue that there could be localized magnetic fields around magnetic minerals on the surface.

Formation of single-domain particles is only the onset of size effects in the nanoworld. If we continue the Democritus experiment and continue to cut the particles into smaller pieces, other size effects start to become apparent (Fig. 1.6). In atoms the electrons occupy discrete energy levels, whereas in a bulk metal the outermost electrons occupy energy bands, in which the energy, for all normal considerations, is a continuum. For nanoparticles smaller than 10 nm, containing about 50,000 atoms, the energy levels of the outermost electrons in the atoms start to display their discrete energies. In other words, the quantum nature of the particles starts to become apparent. In this size range, a lot of the novel and size-dependent behavior can be understood simply in terms of the enhanced proportion of the atoms at the surface of the particles. In a macroscopic piece of metal—for example, a sphere 2 cm across—only a tiny proportion of the atoms, less than 1 in 10 million, are on the surface atomic layer. A 10-nm-diameter particle, however, has 10% of its constituent atoms making up the surface layer, and this proportion increases to 50% for a 2-nm particle. Surface atoms are in a chemical environment that is different from that of the interior and are either exposed to vacuum or interacting with atoms of a matrix in which the

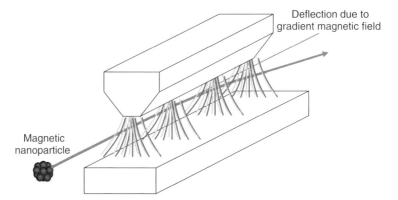

Fig. 1.7 Measuring the magnetic moment in free nanoparticles. The magnetic moment in free nanoparticles can be measured by passing a beam of them through a non-uniform magnetic field and measuring the deflection in their path.

nanoparticle is embedded. Novel behavior of atoms at the surfaces of metals has been known for decades; thus, for example, the atomic structure at the surface is often different from that of a layer in the interior of a bulk crystal. When such a high proportion of atoms comprise the surface, their novel behavior can distort the properties of the whole nanoparticle.

Returning to magnetism, a well-known effect in sufficiently small particles is that not only are they single domains but also the strength of their magnetism per atom is enhanced. A method for measuring the strength of magnetism (or the *magnetic moment*) in small free particles is to form a beam of them (see Chapter 4, Section 4.1.1 for a description of nanoparticle sources) and pass them through a nonuniform magnetic field as shown in Fig. 1.7. The amount the beam is deflected from its original path is a measure of the nanoparticle magnetic moment; and if the number of atoms in the particles is known, then one obtains the magnetic moment per atom.

Magnetic moments of atoms are measured in units called Bohr magnetons[3] or μ_B (after the Nobel laureate Neils Bohr), and the number of Bohr magnetons specifies the strength of the magnetism of a particular type of atom. For example, the magnetic moments of iron, cobalt, nickel, and rhodium atoms within their bulk materials are $2.2\mu_B$, $1.7\mu_B$, $0.6\mu_B$, and $0\mu_B$ (rhodium is a nonmagnetic metal), respectively. Figure 1.8 shows measurements of the magnetic moment per atom in nanoparticles of the above four metals as a function of the number of atoms in the particle. In the case of iron, cobalt, and nickel, a significant increase in the magnetic moment per atom over the bulk value is observed for particles containing less than about 600 atoms. Perhaps most surprisingly, sufficiently

---

[3]A Bohr magneton ($1 \mu_B$) is the magnetic moment of a single free electron produced by its intrinsic angular momentum (spin).

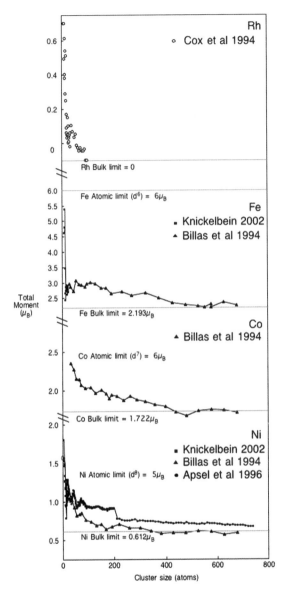

Fig. 1.8   Measured magnetic moments per atom in magnetic nanoparticles. Experimental measurements of the magnetic moment per atom in iron, cobalt, nickel and rhodium (a non-magnetic metal in the bulk) nanoparticles as a function of the number of atoms in the particle. For iron, cobalt and nickel, there is a significant increase in the magnetic moment per atom over the bulk value for particles containing less than about 600 atoms. Rhodium becomes magnetic in particles containing less than about 100 atoms. Note the very dramatic change in the magnetic moment of iron particles in going from a 12-atom particle to a 13-atom particle. Reproduced with the permission of the American Association for the Advancement of Science (AAAS) from I. M. L. Billas et al. [4], permission of the American Physical Society from A. J. Cox et al. [5] and S. Apset et al. [6], Copyright 1994 and 1996 and permission of Elsevier Science from M. B. Knickelbein [7].

small particles (containing less than about 100 atoms) of the nonmagnetic metal rhodium become magnetic.

Throughout the whole size range in Fig. 1.8, the fundamental magnetic behavior of the particles is size-dependent. Do not lose sight of how strange a property this is and how it runs counter to our experience in the macroscopic world. It is as strange as a piece of metal changing color if we cut it in half (something else that happens in nanoparticles). If Democritus were doing his chopping experiment on iron, when he reached a piece 100 nm across, which would be invisible in even the most powerful optical microscope, he would say that he had not yet reached the *atomon* because up to then there would have been no observable change in properties. When he cut in half again, he would suddenly find his piece changing from showing no external magnetism to the full magnetic power of iron with every atomic magnet aligned as the piece formed a single domain particle. He would exclaim "I have reached the *atomon*, let's just try and cut again" (he has nanoscale scissors of course). Imagine his surprise when he finds he can continue and on reaching 3 nm finds that when he cuts again the strength of the magnetism, in proportion to the size of the piece, increases. Is this new piece the *atomon*? Well no, because from then on he would get a change in properties whenever he cut. Some of these changes can be dramatic. For example, if he was holding a 13-atom cluster (let us assume we know what an atom is) and shaved off a tiny piece to produce a 12-atom cluster, the magnetic moment per atom would jump from $2.5\mu_B$ to a staggering $5.5\mu_B$—very close to the single-atom limit of $6\mu_B$.

This highlights one of the most exciting aspects of nanoparticle research. If one considers a nanoparticle as a building block and can assemble large numbers of them to make a material, then it is possible to tailor the fundamental properties of the building block just by changing its size. In Chapter 5 we will look at more sophisticated ways of changing the nanoparticle building blocks. The ability to change the fundamental properties of the building blocks will surely enable us to produce new high-performance materials. As an example, if we deposit iron onto a surface to make a thin film, there is a difference between depositing individual atoms, as with a conventional evaporator, and depositing whole nanoparticles containing, say, 200 atoms. This is clear from Fig. 1.9, which shows a thin film of iron produced by depositing nanoparticles onto a silicon substrate in vacuum. It is clearly a random stack of the deposited particles showing that they have not coalesced to form a smooth film. On this scale a film of the same thickness produced by depositing atoms would appear smooth and featureless, and it would behave differently than the nanoparticle stack.

It is becoming clear that making magnetic thin films out of magnetic nanoparticles may be a way of producing more powerful magnetic materials. First of all, it is worth pointing out that finding a material with an enhanced magnetism per atom over the most magnetic conventional material we have available (iron–cobalt or "permendur" alloy) is regarded as a hugely difficult problem. It has been recognized since 1929 that iron–cobalt alloys have the highest magnetization of all magnetic alloys [9], and materials scientists have been looking for a magnetic

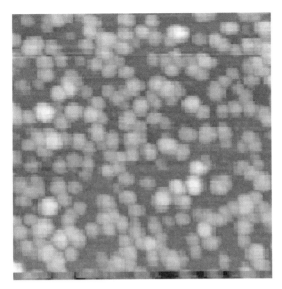

Fig. 1.9 Morphology of nanoparticle film. STM image (see Chapter 4, Section 4.4.1) with an area of 100 nm $\times$ 100 nm of thin film produced by depositing 3 nm diameter iron nanoparticles onto a silicon substrate in vacuum. It is clearly a random stack of the deposited particles and the film properties will be different to those of a smooth film that would be formed by depositing iron atoms. Reproduced with the permission of the American Institute of Physics from M. D. Upward et al. [8].

material with a yet higher performance ever since. Finding such a material would be an important discovery in technology generally. Magnetic materials form a kind of invisible background to our technological existence that we hardly notice, but they are all-pervasive. A useful exercise is to count the number of magnets in one's car, for example. Including all the motors, sensors, and the entertainment system, there will be dozens.

Making high-performance materials is hinted at by the data in Fig. 1.8, which reveal significant increases in magnetism per atom in magnetic nanoparticles. Unfortunately, simply spraying large numbers of iron nanoparticles onto a surface to produce a thick version of the film shown in Fig. 1.9 does not work. The particles in the film would maintain an enhanced magnetic moment per atom but the film is porous, containing a lot of nanoscale voids. The overall performance of the material is a weighted average of the particles and voids; and such a film would have an average magnetic moment per unit volume less than bulk iron, never mind bulk permendur. This is not the end of the story, however, since we can fill the voids by depositing the particles at the same time as a vapor of atoms from a conventional evaporator as shown in Fig. 1.10 [10]. The nanoparticles become embedded in the film (or matrix) produced by the atoms to produce a kind of nanoscale *plum pudding* where the nanoparticles are the plums and the atoms form the pastry. This kind of film is no longer porous and has a density

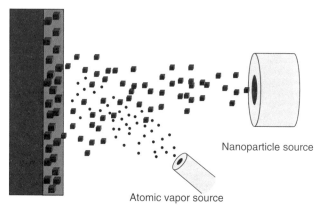

Fig. 1.10   Making granular materials by co-depositing nanoparticles and atoms. A granular material with nanoscale grains can be produced by co-depositing Pre-formed nanoparticles with an atomic vapor from a conventional evaporator. The nanoparticles are embedded in a matrix produced by the atomic vapor to form a granular material in which there is independent control over the volume fraction and grain size.

close to the value found for the bulk materials. If the nanoparticles are iron and the matrix is cobalt (or vice versa), then this method forms a granular version, with nanoscale grains, of the permendur alloy, which has been shown to produce a material with a higher magnetization than conventional permendur [11].

1.3   THE MECHANICAL PROPERTIES OF NANOSTRUCTURED MATERIALS

Mechanical properties such as the strength of metals can also be greatly improved by making them with nanoscale grains. Several basic attributes of materials are involved in defining their mechanical properties. One is strength, which includes characteristics with more precise definitions but basically determines how much a material deforms in response to a force. Others are (a) hardness, which is given by the amount another body such as a ball bearing or diamond is able to penetrate a material, and (b) wear resistance, which is determined by the rate at which a material erodes when in contact with another. These properties are dominated by the grain structure found in metals produced by normal processing. An example of the grains structure of a "normal" piece of metal is shown in Fig. 1.11a, which is an electron microscope image showing the grain structure of tin. Each grain is a single crystal with a typical size of about 20 μm (20,000 nm). The mechanical properties listed above are due to grains slipping past each other or deforming, so clearly what happens at the grain boundaries is very important in determining properties such as strength. By using various techniques including nanoparticle deposition (Fig. 1.10), electrodeposition, and special low-temperature milling methods, it is possible to produce metal samples in which the

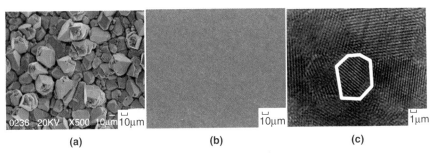

(a)   (b)   (c)

Fig. 1.11   Grain size in nanostructured materials. Electron microscope images showing a comparison of the grain structure in conventional and nanostructured materials. (a)
Conventionally processed material (tin) showing a typical grain size of about 20 μm. (b)
Nanovate™ nanostructured nickel based coating produced by Integran Technologies Inc.
On the same scale as (a) the material appears homogenous. (c) Increasing the magnification by a factor of 15,000 reveals the nano-sized grains. The lines in the picture are
atomic planes and the edges of the grains are revealed by changes in the direction of
the planes as indicated for one of the grains. Reproduced with permission from Integran
Technologies Inc. (http://www.integran.com).

grains are a few nanometers across. An example is shown in Fig. 1.11b, where, on
the same scale as Fig. 1.11a the grain structure disappears to show a homogeneous
material. Blowing up the magnification a further 15,000×, however (Fig. 1.11c),
reveals the new nanoscale grain structure. In this image the individual planes of
atoms are indicated by the sets of parallel lines, and the boundaries are where
the lines suddenly change direction as indicated for one of the grains. Whereas
in the coarse-grained metal shown in Fig. 1.11a, about one atom in 100,000 is
at a grain boundary, in the nano-grained equivalent, about a quarter of the atoms
are at a grain boundary. Clearly this change is going to have a marked effect on
the mechanical properties of the material. Changes in mechanical properties with
grain size were quantified over 50 years ago by Hall and Petch [12, 13], but the
modern ability to vary the grain size right down to the nanometer scale is likely
to yield large increases in performance.

The most dramatic improvement is seen in the "yield strength," which quan-
tifies the load a material can tolerate before it becomes permanently deformed.
All metals are elastic under a small load; that is, when the load is removed, they
return to their original shape, while beyond a certain load they deform plastically
and remain permanently changed. Figure 1.12 [14] shows a plot of the strain
(relative elongation of a sample) versus stress (load) for various nanostructured
aluminum alloys compared to normal (coarse-grained) aluminum alloy. The plastic limit or yield strength occurs at the point where the slope changes, and it is
seen that nanostructured materials have a value that is up to four times higher
that the conventional material. This is a dramatic increase in strength, but even
higher values have been found in other metals—for example, a 10-fold increase
in copper [15]. A problem with nanostructured materials is also revealed by the

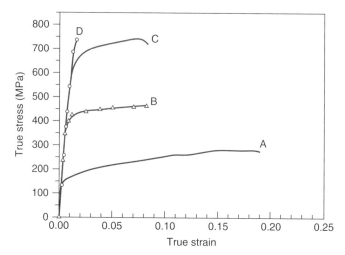

Fig. 1.12   Yield strength of aluminium alloys. Comparison of Deformation (Strain) vs. Load (Stress) for aluminium alloys with different grain sizes. A normal aluminium alloy (coarse-grained). $B - D$ nanostructured aluminium alloy containing grains of size ~30 nm produced by various processes. The plastic limit occurs at the point where the slope changes and the nanostructured materials have a value that is up to four times higher that the conventional alloy. Reproduced with the permission of Elsevier Science from K. M. Youssef et al. [14].

plot, however; that is, they fail (break) at relatively low strains. Problems such as this are being addressed by improvements in processing [15].

1.4   THE CHEMICAL PROPERTIES OF NANOPARTICLES

Another size-dependent property of nanoparticles is their chemical reactivity. This is demonstrated most dramatically by gold, which in the bulk is the archetypal inert material. This is one of the reasons why it is so highly valued since it does not corrode or tarnish and thus has a timeless quality. The only acid that is known to attack it is a hellish brew of concentrated nitric and hydrochloric acids mixed to form what has been poetically named aqua regia (royal water). It would therefore seem that gold, would be useless as a catalyst to speed up chemical reactions, but this is not so for gold nanoparticles. In fact, most catalysts are in the form of nanoparticles; and bearing in mind everything that has been said so far about how material properties change with size in the nanoworld, the same is true of gold. When gold is in the form of nanoparticles with diameters less than about 5 nm, it becomes a powerful catalyst, especially for the oxidation of carbon monoxide (CO). The full story is quite complicated because the reactivity of gold nanoparticles appears to depend not only on their size but also on the material on which they are supported.

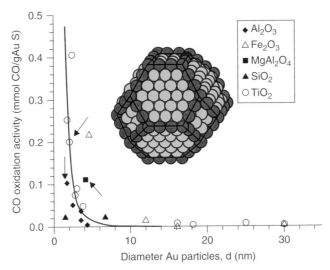

Fig. 1.13 Reactivity of gold nanoparticles. Measured activities of gold nanoparticles on various supports (box) for carbon monoxide oxidation as a function of particle size. The black line is a fit using a $1/d^3$ law and is seen to broadly represent the variation indicating that the dominant effect size effect is the proportion of gold atoms that are at a corner between facets at the surface (see text). Such atoms are highlighted in red on the nanoparticle shown. Reproduced with the permission of Elsevier Science from N. Lopez et al. [16].

A recent assessment of a number of research papers on the effectiveness of gold in catalyzing the above reaction, however [16], has concluded that the dominant effect is that of the gold nanoparticle size, with the nature of the support playing a secondary role. Figure 1.13 shows a compilation of data on the carbon monoxide (CO) oxidation activity of gold nanoparticles on various supports as a function of their size and shows the impressive performance of the gold, which is completely inert in macroscopic-sized pieces. Since catalysis can only occur at the surface layer of atoms, the dominant size effect is the proportion of gold atoms that are at the surface. In fact the most important atoms for catalysis are those at the corners between different facets. These low coordinated atoms are where the reacting carbon monoxide (CO) molecules preferentially bond to during the reaction. The fraction of this type of atom (highlighted in red in Fig. 1.13) is proportional to $1/d^3$, where d is the particle diameter and the black line in Fig. 1.13 is a fit to the data using this law demonstrating that the dominant size effect is indeed the proportion of corner atoms at the surface. The focus here has been on gold because of its extreme demonstration of size-dependence—that is, from a completely inert material to a powerful catalyst—but as a general rule the performance of all catalysts depends on the particle size. Because of the importance of catalysts to the chemical industry, the effect is the focus of a large amount of research activity.

1.5 NANOPARTICLES INTERACTING WITH LIVING SYSTEMS

The novel properties of nanoparticles—for example, the enhanced chemical reactivity—makes their interaction with biological systems dependent on particle size. This can be illustrated with the example of silver nanoparticles. It has been known for some time that silver is highly toxic to a wide range of bacteria, and silver-based compounds have been used extensively in bactericidal applications. This property of silver has caused great interest, especially because new resistant strains of bacteria have become a serious problem in public health. For example, MRSA bacteria kill 5000 hospital patients a year in the United Kingdom alone; and any method of attacking them, not involving normal antibiotics, is becoming increasingly important. Trials of silver-loaded wound dressings and special fabric with silver yarn woven into it have proved highly successful in treating and preventing the spread of MRSA infections [17]. There is also interest from less urgent but nevertheless important applications such as slow release of silver from compounds for preservatives or long-lasting bactericidal protection in plastics.

Silver in the form of nanoparticles becomes yet more effective, partly because of the high surface/volume fraction so that a large proportion of silver atoms are in direct contact with their environment. In addition, nanoparticles are sufficiently small to pass through outer cell membranes and enter cells' inner mechanisms. A recent study using nanoparticles with a wide size range [18] showed that only silver nanoparticles with sizes in the range 1–10 nm were able to enter cells and disrupt them. They were found to do this in two ways. Some particles attached to the cell membrane and disturbed its functions, such as respiration. Others penetrated the outer membrane and caused further damage, including damage to the cell DNA. Figure 1.14 shows an electron microscope image of some P. *aeruginosa*[4] bacteria after they had been exposed to silver nanoparticles, and the high magnification inset reveals some nanoparticles attached to the cell membranes. This strain can be particularly problematic to cystic fibrosis sufferers because it readily colonizes the favorable environment found in their lungs. It protects itself by producing a thick mucus layer, which makes it difficult to treat with antibiotics once it is established. Silver nanoparticles suspended in an aerosol could prove to be an effective treatment.

There may also be some role for nanoparticles in the fight against AIDS. The same team that carried out the work described above also exposed the HIV-1 virus to silver nanoparticles of the same size [19]. There has been very little work on the interaction of nanoparticles with viruses, which are much smaller than bacteria and are themselves inhabitants of the nanoworld (see Fig. 0.1). It was discovered that, again, the important size range is 1–10 nm; and, as shown in Fig. 1.15, nanoparticles of this size attached themselves in an ordered array on the surface of the virus. The HIV-1 virus is covered in a regular arrangement

---

[4]Bacterium that infects the lungs of cystic fibrosis sufferers and causes inflammation and breathing problems.

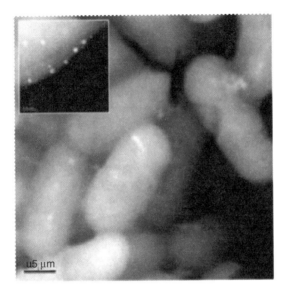

Fig. 1.14 Silver nanoparticles attacking bacteria. Electron microscope image of P.aeruginosa[4] bacteria after exposure to silver nanoparticles. The particles with size ranges 1–10 nm are active in disrupting the cell function. The inset shows a high-resolution image of some particles attached to the cell outer membrane. Reproduced with the permission of the Institute of Physics from J. R. Morones et al. [18].

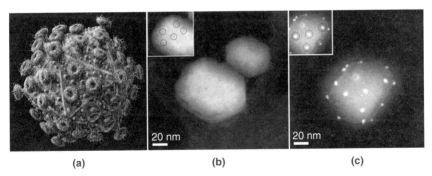

     (a)   (b)   (c)

Fig. 1.15 Silver nanoparticles attached to HIV-1 virus. (a) Computer generated image of HIV virus, showing glycoprotein 'knobs' (Reproduced with permission from www.virology.net/Big_Virology/BVretro.html). (b) Electron microscope image, with inset superimposing position of knobs. (c) Electron microscope image of virus after attachment of silver nanoparticles. Reproduced with the permission of BioMed Central from J. L. Elechoguerra et al [19].

of glycoprotein "knobs" (Fig. 1.15), and the average distance between the silver nanoparticles in the image suggests that they attach themselves to these knobs. The team went on to demonstrate that the nanoparticles prevent the virus from binding to host cells.

Here we have only encountered very brief sketches of a couple of examples of applications of nanoparticles in health care. This is a huge and burgeoning field, and a more comprehensive discussion is postponed until Chapter 6. As a final example in this chapter, it is worth noting that there is increasing use of nanoparticles in cosmetics such as face creams. In order to replenish certain compounds in the skin, whose reduced concentration is partly responsible for the effects of aging, it is necessary to enable these compounds to penetrate the outermost layers of skin. This can mean either producing the compounds themselves as nanoparticles or binding them onto "carrier" particles that transport them to the deeper layers. Some sunscreen manufacturers now use titanium dioxide nanoparticles in new products as the carriers.

The key aims of this chapter have been to give an impression of the nanometer size scale and to introduce the idea that sufficiently small pieces of matter behave differently than the bulk material. In addition, the novel behavior of nanoparticles depends on their size, and several examples have been introduced where the unusual behavior can be put to good use in technology. All the topics discussed fall into the category of incremental nanotechnology, and it is evident that this category is already being used or will shortly be used in industrial sectors ranging from magnetic recording to medicine. All the types of nanoparticle presented here are artificially produced; but in the following chapter, naturally occurring nanoparticles and their effect on the environment and our health will be discussed.

PROBLEMS

1. Silver, whose density is 10,490 kg/m^3, is used to form spheres of different diameters. Calculate the surface area in m^2 of 1 g of Ag spheres of diameter:

(a) 1 mm

(b) 10 μm

(c) 10 nm

2. Calculate the proportion of atoms on the surface atomic layer of Ag spheres with the diameters below given that the thickness of the surface atomic shell is 0.289 nm, stating any assumptions made.

(a) 1 μm

(b) 10 nm

(c) 5 nm

3. In a magnetic material the energy of the exchange interaction between neighboring atoms is about 1 eV/atom, whereas the magnetic dipolar interaction between neighboring assemblies of atoms has an average value of 10^{-5}

eV/atom. Use this data to estimate the maximum size of a spherical single-domain particle. (*Hint*: Assume that the exchange interaction only operates between atomic neighbours, whereas in a small particle the dipolar interaction affects all atoms equally. In addition, for an estimate, assume that the atoms are cubes with a width of 0.2 nm.[5])

4. An HIV virus with a diameter of 100 nm has 50 glycoprotein extrusions over its surface, each one of which will bind a 5-nm-diameter Ag nanoparticle. You are provided with a suspension of 5-nm-diameter Ag nanoparticles containing 1 μg Ag per milliliter of liquid. What volume of the suspension is required to saturate 10^7 viruses in a test tube?

REFERENCES

1. M. D. Upward, P. Moriarty, P. H. Beton, S. H. Baker, C. Binns, and K. Edmonds, Measurement & manipulation of Mn clusters on clean and fullerene terminated Si(111)-7×7, *Appl. Phys. Lett.*, **70** (1997), 2114.

2. D. Schüler, The biomineralisation of magnetosomes in *Magnetospirillium gryphiswaldense*, *Int. Microbiol*. **5** (2002), 209–214.

3. D. S. McKay, E. K. Gibson, Jr., K. L. Thomas-Keprta, H. Vali, C. S. Romanek, S. J. Clemett, X. D. F. Chillier, C. R. Maechling, and R. N. Zare, Search for past life on Mars: Possible relic biogenic activity in Martian meteorite ALH84001, *Science* **273** (1996), 924–930.

4. I. M. L. Billas, A. Chatalain, and W. A. de Heer, Magnetism from the Atom to the Bulk in Iron, Cobalt, and Nickel Clusters, *Science* **265** (1994), 1682.

5. A. J. Cox, J. G. Louderback, S. E. Apsel, and L. A. Bloomfield, Magnetism in 4d-transition metal clusters, *Phys. Rev. B* **49** (1994), 12295–12298.

6. S. E. Apsel, J. W. Emmert, J. Deng, and L. A. Bloomfield, Surface-enhanced magnetism in nickel clusters, *Phys. Rev. Lett*. **76** (1996), 1441–1444.

7. M. B. Knickelbein, Adsorbate-induced enhancement of the magnetic moments of iron clusters, *Chem. Phys. Lett*. **353** (2002), 221–225.

8. M. D. Upward, B. N. Cotier, P. Moriarty, P. H. Beton, S. H. Baker, C. Binns and K. Edmonds, Deposition of Fe clusters on Si surfaces, *J. Vacuum Sci. Technol. B* **18** (2000), 2646–2949.

9. T. Sourmail, Near equiatomic FeCo alloys: Constitution, mechanical and magnetic properties, *Prog. Mater. Sci*. **50** (2005), 816–880.

10. P. Melinon, V. Paillard, V. Dupuis, A. Perez, P. Jensen, A. Hoareau, J. P. Perez, J. Tuaillon, M. Broyer, J. L. Vialle, M. Pellarin, B. Baguenard, and J. Lerme, From free clusters to cluster-assembled materials, *Int. J. Modern Phys*. **B9** (1995), 339–397.

---

[5]You should find that your estimate is much larger than the critical size (~100 nm) for a single domain particle given in the text. The reason is that the exchange energy in a domain boundary can be reduced by a large factor by spreading it over a number of atomic layers. That is, instead of having an abrupt $180°$ reversal of the magnetization across a single atomic plane, the magnetization rotates a fraction of $180°$ across each plane. This brings the critical size down to ~100 nm).

11. C. Binns, S. Louch, S. H. Baker, K. W. Edmonds, M. J. Maher, and S. C. Thornton, High-moment films produced by assembling nanoclusters, *IEEE Trans. Magn.* **38** (2002), 141–145.

12. E. O. Hall, The deformation and ageing of mild steel: III Discussion of results, *Proc. Phys. Soc. B* **64** (1951), 747–753.

13. N. J. Petch, *J. Iron Steel Inst.* (May 1953), 25.

14. K. M. Youssef, R. O. Scattergood, K. L. Murty and C. C. Koch, Nanocrystalline Al–Mg alloy with ultrahigh strength and good ductility, *Scr. Mater.* **54** (2006), 251–256.

15. C. C. Koch, K. M. Youssef, R. O. Scattergood, and K. L. Murty, Breakthroughs in optimization of mechanical properties of nanostructured metals and alloys, *Adv. Eng. Mater.* **7** (2005), 787–794.

16. N. Lopez, T. V. W. Janssens, B. S. Clausen, Y. Xu, M. Mavrikakis, T. Bligaard, and J. K. Norskov, On the origin of the catalytic activity of gold nanoparticles for low-temperature CO oxidation, *J. Catal.* **223** (2004), 232–235.

17. BBC News January 17, 2006 (www.bbc.co.uk).

18. J. R. Morones, J. L. Elechiguerra, A. Camacho, K. Holt, J. B. Kouri, J. T. Ramirez, and M. J. Yacaman, The bactericidal effect of silver nanoparticles, *Nanotechnology* **16** (2005), 2346–2353.

19. J. L. Elechoguerra, J. L. Burt, J. R. Morones, A. Camacho-Bragado, X. Gao, H. H. Lara, and M. J. Yacaman, Interaction of silver nanoparticles with HIV-1, *J. Nanobiotechnol.* **3** (2005), 1–10.

Nanoparticles Everywhere

2.1 NANOPARTICLES IN THE ATMOSPHERE

In the previous chapter, the special properties of pieces of matter with nanoscale dimensions were presented and how this novel behavior could be exploited in technologies as diverse as advanced engineering materials and health care. This chapter is about naturally occurring nanoparticles, both in our immediate environment—that is, the Earth's crust, oceans, and atmosphere—and out into deep space, where nanoparticles with their unique properties have helped shape the observable universe. The particles in the Earth's atmosphere have an important influence on the climate but also have a poorly understood effect on life and our health. Improving our understanding of the effect of airborne nanoparticles is becoming increasingly important in a world where nanotechnology is poised to become a major activity. Clearly, the amount of manufactured nanoparticles will increase, so it is wise to be aware of how they interact with life and with the environment. It is important to emphasize, however, that manufactured nanoparticles are normally bound up in some material and that the number of "loose" particles produced by nanotechnology will not necessarily become significant compared to those produced by the natural processes described below. In this chapter the discussion is extended to encompass nanoparticles that are generated by existing human activities not directly involving nanotechnology, such as power generation, transport, and so on. Obviously, these are not naturally occurring in the normal sense of the phrase, but they are a component of a pre-nanotechnology background of nanoparticles in which we live. The effect of naturally occurring nanoparticles on the environment is an enormous multidisciplinary subject, and a rigorous discussion is well beyond the scope of this book. It is an important hot topic, however, because it encompasses climate change, and nanoparticles are implicated in many of the feedback mechanisms involved in the Gaia hypothesis

Introduction to Nanoscience and Nanotechnology, by Chris Binns
Copyright © 2010 John Wiley & Sons, Inc.

that treats the Earth as a living organism. The aim of this chapter is to describe in general terms where the nanoparticles come from and, as in the previous chapter, emphasize the special nature of particles belonging to the nanoworld (<100 nm; see Fig. 0.1).

Naturally occurring nanoparticles are ubiquitous in land, sea, and air and come from a number of processes (Fig. 2.1), including volcanic activity, forest fires, ocean bed hydrothermal vents,[1] geological processes, living creatures, and human industrial activity. They are also to be found in space (though normally called dust by astronomers) and produced by, among other things, supernova explosions.[2]

To begin with, we will focus on atmospheric nanoparticles because these probably have the most immediate effect on living things. The general term for tiny (solid or liquid) particles suspended in a gas is an *aerosol*. This was first used in the 1920s to distinguish air-suspended particles from liquid suspensions, or hydrosols. The term *suspension* implies that the particles are defying gravity, but this of course is not the case. The particles are falling through the gas (a viscous medium), but their terminal velocity due to gravity is so low that it may take years for them to settle (see Advanced Reading Box 2.1). In this regime, for all practical purposes we can consider them to be suspended. Other processes can, however, remove nanoparticles from an aerosol. To begin with, if they have a sufficiently high concentration, they will agglomerate and the larger particles will settle much more rapidly. In addition, in a humid atmosphere, nanoparticles will act as nuclei for the formation of water droplets (see below); and if these grow large enough to fall as rain, this will act as a removal mechanism for nanoparticles ("rainout"). Alternatively, the particles can be incorporated into existing raindrops and removed ("washout").

Until recently, naturally occurring nanoparticles in the atmosphere have been relatively overlooked because most natural processes that generate particles produce a wide size range that encompasses chunks up to macroscopic dimensions. Within such a wide distribution, the mass fraction (or volume fraction) of particles belonging to the nanoworld (<100 nm) is a tiny proportion of the whole distribution. For many processes in which the particles are interacting with their environment, however, it is the number density that is the important parameter, and here the nanoparticles dominate. Figure 2.2 shows the concentration of particles as a function of their diameter in a typical urban aerosol using three different measures. The lower curve shows the total volume of particles in cubic micrometer per cubic centimeter of air. In this way of measuring the aerosol

---

[1]Ocean bed hydrothermal vents are fissures on the sea floor from which geologically heated water erupts into the surrounding cold water. They are mineral-rich and support a variety of deep-sea creatures.

[2]Supernovae are massive stellar explosion that occur at the end of the life of massive stars. After the star exhausts its supply of hydrogen, it collapses under gravity and if it is above a critical mass (several times that of our sun); the resulting implosion at the core produces a thermonuclear explosion, which for a period of weeks can produce billions of times more power than a normal star.

concentration, it would appear that particles with sizes in the range 0.5–10 μm dominate the distribution. These tend to be mechanically generated—for example, tyre dust, wind-blown sand grains, and so on. In contrast, the upper graph shows the distribution when we measure simply the number of particles per cubic centimeter. The distribution is almost entirely in the nanoworld and dominated by nanoparticles with a typical size of 10–20 nm, with larger particles being virtually absent on the same scale. These are mostly produced by combustion sources or by chemical reactions resulting in nitrates and sulfates. Some are smaller sea-salt crystals produced by a bubble-bursting mechanism, described below, and carried over land by air currents. In situations where *each particle* does something—for example, in respiratory problems—this upper graph is the relevant distribution. Note that according to the figure, each lungful of air in the urban environment contains millions of nanoparticles. This is a figure to bear in mind when we talk of the hazards of nanoparticles resulting from nanotechnology. It is hard to imagine nanotech industries producing loose aerosol over the world on this scale.

ADVANCED READING BOX 2.1—TERMINAL VELOCITY OF AEROSOL PARTICLES

It is easy to show [1] that for large (micron-sized or more) particles with a diameter d and a density ρ_p, their terminal velocity due to gravity in a still gas with a density ρ_g is

$$V_t = \frac{(\rho_p - \rho_g)d^2 g}{18\eta},\qquad (2.1)$$

where η is the viscosity of the gas ($\eta = 1.81 \times 10^{-5}$ Pa·s for air at standard conditions). For 1-μm-diameter particles with a typical density (1000–5000 kg/m^3), this gives $\sim$0.1 mm/sec. The equation, however, is only valid for relatively large particles. In its derivation it is assumed that the gas velocity at the particle surface is zero, which is not valid for very small particles whose size is less than the mean free path of the gas molecules. To put it crudely, very small particles "slip" through the gaps between the gas molecules and fall faster than predicted by the equation. As the particles get smaller, an increasing slip correction factor needs to be applied; this can get to be a factor of 10 or more. Even so, the fact that the terminal velocity decreases as d^2 ensures that small particles do drop more slowly. Applying the slip correction factor to 10 nm-diameter particles falling through air gives a terminal velocity of $\sim$0.1 μm/sec (or about a meter every four months). For all practical purposes, nanoparticles can be assumed to be *suspended* in the atmosphere. This discussion, however, is only relevant to settling of particles by gravity. Accumulation, rainout, and washout will remove them more rapidly.

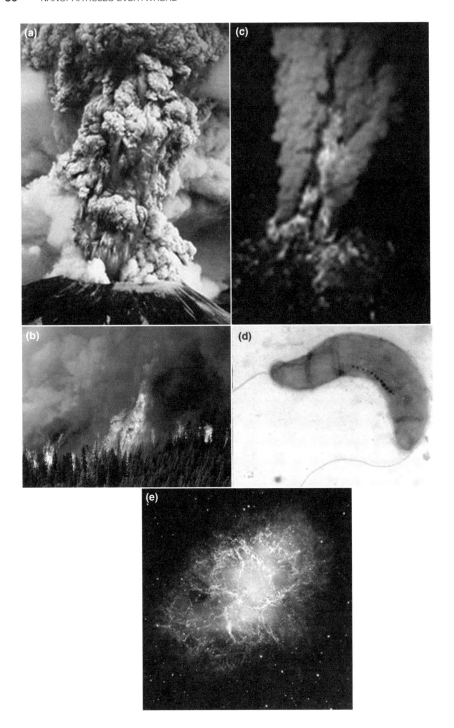

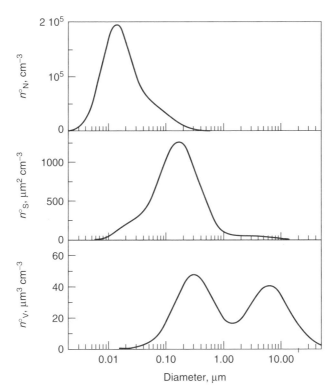

Fig. 2.2   Size distribution of urban aerosol. The concentration of airborne particles (aerosol) as a function of their diameter in a typical urban environment using three different measures. The lower curve shows the total volume of particles in cubic μm per cubic centimeter of air. The middle curve shows the total surface area of the particles in square μm per cubic centimeter of air. The upper curve gives simply the number of particles per cubic centimeter of air. It is evident that while the total volume (and also the mass) contained in the particles per cubic centimeter is concentrated in large particles outside of the nanoworld. In terms of particle numbers the distribution is completely dominated by nanoparticles with a typical size of 10–20 nm diameter. Reproduced with the permission of Wiley from J. H. Seinfeld and S. N. Pandis [2].

←

Fig. 2.1   Sources of background nanoparticles. (a) Volcano and (b) forest fires produce nanoparticles in the atmosphere (aerosols). (c) Hydrothermal vents produce nanoparticles in the ocean (hydrosols). (d) Some bacteria produce nanoparticles such as this river-dwelling bacterium that manufactures magnetic nanoparticles (black dots). (e) Supernova explosions such as the crab nebula shown here spread nanoparticles through space. (a) Reproduced with permission from Science museum webpage: (http://www.sciencemuseum.org.uk/antenna/nano/planet/143.asp) (b) Reproduced with permission from government of British Columbia. (c) Reproduced from Wikipedia web page: (http://en.wikipedia.org/wiki/Hydrothermal_vent) (d) Reproduced with the permission of the Spanish Society for Microbiology from D. Schüler [2]. (e) Reproduced from Wikipedia web page: (http://en.wikipedia.org/wiki/Supernova).

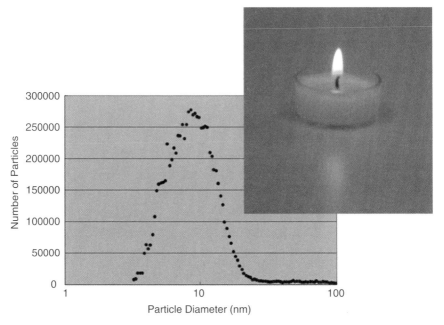

Fig. 2.3 Nanoparticles produced by candles. Size distribution of particles produced by combustion from a standard 'nightlite' type candle. The distribution was measured using an aerosol particle sizer and if plotted as the number density, consists mostly of nanoparticles with diameters around 10 nm. These are predominantly carbon particles. The method used to measure the size distribution is described in Chapter 4, Section 4.1.6.

A common process that produces nanoparticles in the atmosphere is gas-to-particle conversion (GPC). When a vapor is rapidly cooled—for example, in combustion where hot gases meet cool air—the atoms or molecules in the vapor condense to form particles. Alternatively, a vapor produced by some process may chemically react with the atmosphere to produce a less volatile compound, which then condenses into solid or liquid particles. Just about any combustion will generate a cloud of nanoparticles. For example, Fig. 2.3 shows the particles generated by a "nightlite"-type candle. Burning petrol and diesel also generates nanoparticles, so in urban areas this is a major source. A particular issue is with diesel engines, which generate most of their nanoparticles from sulfur contained in the fuel. There is a worldwide effort to reduce the levels of sulfur in diesel and thus the number of nanoparticles produced.

2.2 ATMOSPHERIC NANOPARTICLES AND HEALTH

The effect on the body of exposure to nanoparticles is a hot topic in the debate on the benefits/hazards of nanotechnology, but there is limited hard information to inform the discussion. It is worth emphasizing that life on this planet evolved in a dense cloud of naturally occurring nanoparticles, and we have been exposed to nanoparticles produced by human activity since the invention of fire. In addition, industrial activity based on certain types of nanoparticle—for example, carbon black—has been around since early civilizations. The difference now is that new nanotechnology industries will produce nanoparticles composed of a wide range of different materials as well as other forms of "nanomaterial" such as carbon nanotubes (see Chapter 3).

There are three general surfaces in the body in contact with the environment—that is, the skin, the lungs, and the intestines. The skin, with a typical surface area of 1.5 m^2, is, in principle, a barrier, whereas the lungs and intestines are designed to exchange material with the environment. Within the lungs, a system of finer and finer tubes terminates at clusters of tiny air sacs called alveoli (Fig. 2.4a), typically 200 μm (0.2 mm) across. There is capacity for a huge number of these, and an adult lung contains approximately 500 million alveoli, thus presenting an area of about 100 m^2 (about half a tennis court) to the environment. The alveoli are encased in blood vessels, and the barrier separating the flowing blood and air in the alveoli is about 300 nm, which is thin enough to allow gas molecules to diffuse through. Particles that enter the lungs, if they are smaller than a few microns, can pass right through the tubular system and into the alveoli. Nanoparticles, with diameters less than 100 nm, almost all end up in the alveoli.

Since humans evolved in the presence of atmospheric aerosol, the body has well-developed mechanisms to cope with inhaled particles including nanoparticles. The primary mechanism that removes particles from the lungs involves *macrophages* (Fig. 2.4b) that ingest particles using a mechanism called *phagocytosis* in which the cell wall of the macrophage envelops the foreign body and passes it through to the interior. These specialized cells, with typical sizes of 10–20 μm, inhabit the lungs right down to the alveoli. They are passed, with their cargo of absorbed particles, to the top of the lungs by a gradual upward drift of mucus, called the mucociliary escalator, so they can be swallowed and passed out through the digestive system. It typically takes the escalator about 70 days to clear a particle from the lungs, though this time increases if the number of particles starts to overwhelm the alveolar macrophages.

The system copes admirably with the huge number of particles that we constantly breathe in, but there is no doubt that it struggles with nanoparticles. It has been known for some time that phagocytosis works best for particles with sizes around 1 μm but becomes very inefficient for particle sizes below 100 nm [3]. More recent studies have shown that the clearance time for nanoparticles of titanium dioxide with a diameter of 20 nm is significantly longer than particles with a diameter of 200 nm [4]. In addition, titanium dioxide in the form of nanoparticles results in a greater incidence of tumors than do larger particles. Other studies have shown that pre-exposure of macrophages to titanium dioxide or carbon nanoparticles limits their ability to subsequently ingest larger particles by phagocytosis [5]. A condition caused by inhaled nanoparticles familiar with welders is "fume fever." Welding can produce large quantities of metal or oxide nanoparticles,

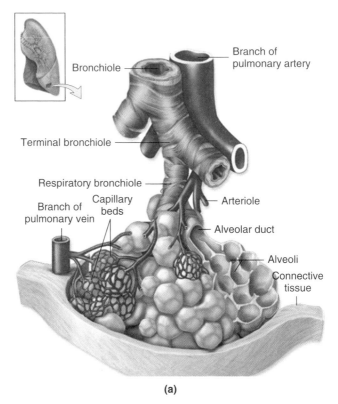

Bronchiole

Branch of pulmonary artery

Terminal bronchiole

Respiratory bronchiole

Branch of pulmonary vein

Capillary beds

Arteriole

Alveolar duct

Alveoli

Connective tissue

(a)

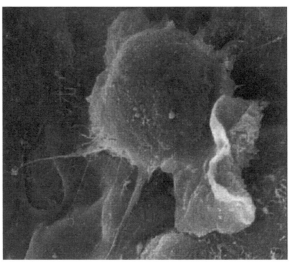

(b)

and overexposure to these can cause an allergic reaction with flu-like systems. A review of the toxicology of inhaled particles derived from combustion, including vehicle exhausts, has been compiled by Donaldson et al. [6].

The skin consists of three layers called the epidermis, the dermis, and the subcutaneous layer and is in principle a barrier to the environment including particles. The interaction of particles of titanium dioxide with the skin has been intensively studied because these are often used in sunscreens to absorb UV light. Micron-sized particles have limited penetration of the epidermis but can reach the dermis. Nanoparticles of titanium dioxide with sizes in the range 5–20 nm easily penetrate into the skin and can interact with the immune system [7].

Particles can also be taken up by the intestines, whose highly convoluted interior provides a surface area in the region of 200 m^2—that is, even more than the lungs. The detailed mechanisms are beyond the scope of this book, but an excellent review has been compiled by Hussain et al. [8].

It is clear that more research on how nanoparticles interact with the body needs to be done before any accurate assessment of the hazards of nanoparticles produced by nanotechnology can be made. An assessment of the presently known health effects of nanoparticles has been provided by the Health and Safety Executive [9]. Current air policies on dust levels do not distinguish particle sizes except in a very broad-brush manner and focus on all particles smaller than 10 μm (the so-called PM_{10} fraction). The United States has introduced an additional air quality standard for $PM_{2.5}$ for particles smaller than 2.5 μm—the so-called fine particle fraction with the European Union considering setting similar limits. It is clear from the previous discussion that in the future there will need to be further limits set at $PM_{0.1}$ and probably $PM_{0.05}$ (particles smaller than 50 nm).

2.3 NANOPARTICLES AND CLIMATE

The presence of aerosol in the atmosphere (and it is worth emphasizing again that when measured as the number of particles per unit volume, nanoparticles are the dominant component) is hugely important for our environment. It is not overstating the point that it is responsible for the climate as we know it.

Fig. 2.4 Nanoparticles and the lungs. (a) Lung structure showing the ends of the finest bronchial tubes terminating in clusters of tiny sacs (alveoli) 0.2 mm across and wrapped in blood vessels. The adult lung contains about 500 million of these with a total surface area of about 100 m^2. Across a significant proportion of this surface the tissue separating air and blood is as thin as 300 nm and allows gas molecules to diffuse across. Reproduced with permission of McGraw-Hill from *Human Anatomy* by McKinley and O'Loughlin, 2006. (b) Alveolar macrophages, typically 1–2 μm across that occupy the lungs including the alveoli. They ingest particles by a process called phagocytosis and are carried up through the lung system by a slow mucus flow called the mucociliary escalator. They are eventually swallowed and pass out through the digestive system.

The most important contribution of the aerosol is its role in the formation of clouds. Pure water vapor in the atmosphere is invisible; but when it condenses into microscopic water droplets (suspended as an aerosol) over a large region of sky, a cloud is born. The process of cloud formation and how they evolve and precipitate is a huge subject; but an important fundamental consideration relevant to this book is that without a preexisting aerosol of particles, clouds would not form. In a purely gaseous atmosphere, even one saturated with water vapor, it is highly improbable for water droplets to start growing, unless there are some initial "seed" particles that water can condense onto. These seeds are referred to as cloud condensation nuclei (CCNs). The reason why pure water vapor will not form droplets is described briefly in Advanced Reading Box 2.2; but in a nutshell, although water molecules do stick together, at normal temperatures and vapor pressures, they do not stay together long enough for a third and fourth molecule to join them and start a droplet growing. If a water droplet above a critical size (just a few molecules) were somehow created, it would be stable and in a humid atmosphere would grow. Without CCNs, however, there is no way to achieve a water droplet of the critical size. The presence of preexisting CCNs changes that, and water molecules can easily condense onto them and grow to a normal cloud droplet size. These fall sufficiently slowly under gravity to be considered as suspended (see Advanced Reading Box 2.1). Under certain conditions the cloud droplet can grow large enough to drop out of the cloud as rain. These raindrops of course contain the CCNs that started the water drop growing in the first place; so although there is a tendency to regard rainwater as pure, it contains the particles that formed the original CCNs. If these contain sulfur, the rain will be acidic to a degree and as described below there are natural processes that produce sulfur-containing aerosol, so a certain amount of acid rain is inherent in climate processes and has nothing to do with human activities. The relative sizes of CCNs (mostly nanoparticles), cloud droplets, and raindrops are illustrated in Fig. 2.5. Precipitating clouds are a mechanism for removing atmospheric aerosol and thus form a self-regulating feedback system. An increase in the density of aerosol produces more CCNs, which produce more cloud, which in turn increases the rate at which particles are washed out back to the ground.

ADVANCED READING BOX 2.2—CONDENSATION OF WATER DROPLETS IN A HUMID ATMOSPHERE

The vapor pressure above a flat liquid surface within a closed container is [10]

$$p_0 = \frac{\pi n_s}{\theta} kTe^{-E_f/kT}, \tag{2.2}$$

where n_s is the atomic density near the surface, E_f is the enthalpy of evaporation (or the energy required by a molecule to escape from the flat surface), θ is the sticking probability of a vapour phase molecule incident on the liquid surface, k is Boltzmann's constant, and T is the temperature. Since the term kT varies slowly compared to the exponential term, for the present purposes, (2.2) can be simplified to

$$p_0 = Ae^{-E_f/kT}, \qquad (2.3)$$

where A includes all the constants in (2.2). If we now consider a curved liquid surface—say a drop, in equilibrium with its vapor—a molecule near the surface has, on average, slightly fewer nearest neighbors because of the curvature. As a result, the enthalpy will decrease and the vapor pressure will be greater than above the flat surface. The enthalpy becomes dependent on the radius of the drop, and it can be shown [10] that the enthalpy is

$$E_c(r) = E_f - \frac{2\gamma v}{r}, \qquad (2.4)$$

where $E_c(r)$ is the radius-dependent enthalpy for a curved surface, γ is the surface tension of the drop, and v is the volume of the departing molecule. This equation is derived by working out how much the surface energy of a drop changes as a result of losing a molecule. The increased vapor pressure, $p > p_0$ of a drop compared to a flat surface is obtained by replacing E_f in equation (2.3) with $E_c(r)$ given by (2.4), that is,

$$p = p_0 e^{2\gamma v/rkT}. \qquad (2.5)$$

So now we can consider what happens if we have a vapor with a pressure $p > p_0$ (a supersaturated vapor) containing no liquid drops. If we introduce a drop with a radius r derived from (2.5) into this vapor, it will be stable because the rate of molecules evaporating from it will equal the rate of molecules incident on it from the vapor. If our initial drop is smaller than r, however, it will shrink because it will evaporate molecules faster than acquiring them from the vapor. Similarly, a larger initial drop will grow. In a highly pure vapor, getting the initial stable size drops is a bottleneck because the only way they can form is by the simultaneous collision of a sufficient number of molecules (homogenous nucleation), which is a highly improbable event. If there are, however, preexisting particles (liquid or solid) in the supersaturated vapor, it quickly condenses onto these. In the case of clouds, these preexisting particles are called cloud condensation nuclei or CCNs.

CCNs are an example of where it is the number density of particles that is important rather than the mass they contain. Each particle will act as a perfectly good CCN whether it is a tiny nanoparticle or a micron-sized particle (though the character of clouds seeded by different sized particles

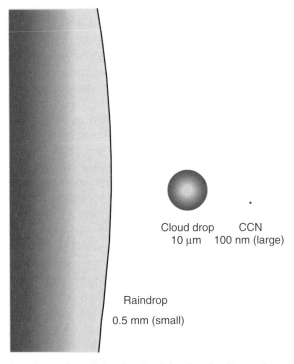

Cloud drop   CCN
10 μm   100 nm (large)

Raindrop

0.5 mm (small)

Fig. 2.5   Relative sizes of particles involved in clouds. Comparison of a cloud condensation nucleus (CCN), a typical water droplet in a cloud and a raindrop. In order to get a meaningful picture a large CCN has been compared to a small raindrop. CCN's can be a factor of 10 smaller and raindrops can be a factor of 10 larger. The CCN's are the pre-existing particles that allow cloud droplets to form (see Advanced Reading Box 2.2). Normal cloud drops fall sufficiently slowly under gravity to be 'suspended' (see advanced reading box 2.1) but under certain circumstances can grow large enough to precipitate out.

may be different). So most CCNs are in fact nanoparticles, and the special properties of matter at the nanoscale come into play. For example, the growth of cloud droplets is profoundly affected if the CCNs are soluble in water. One mechanism is that soluble CCNs can change the surface tension of the water droplets condensing onto them and thus change the stable droplet size for a given water vapor pressure (see Advanced Reading Box 2.2). The fact that CCNs are nanoparticles is important since their solubility becomes dependent on their size, and substances that are insoluble in the bulk or as coarse particles can become soluble in sufficiently small particles. Even quartz, which is the archetypal insoluble material (fortunate for those of us who enjoy a glass of beer), becomes soluble at particle sizes of a few nanometers.

2.4   MARINE AEROSOL

A significant proportion of atmospheric nanoparticles are generated above the oceans. These are known as *marine aerosol* and are produced by a number of sources. The simplest to understand are sea-salt particles, which are produced when bursting bubbles at the surface produce a spray of droplets of brine from which the water evaporates to leave salt particles. These have a wide size range, but all are small enough to form an aerosol; and as with most aerosols, when measured as the number of particles per unit volume, the nanoparticles dominate, with most particles having a size around 30 nm [11]. Bearing in mind how slowly these fall out due to gravity, they are easily carried into all levels of the atmosphere by winds and updrafhts and a significant proportion of the aerosol over land is sea-salt particles.

The story of marine aerosol becomes much more complicated when life is included, since plankton and microorganisms on the ocean surface enable other mechanisms for producing particles. A common example starts with the chemical dimethyl sulfide (DMS), which is produced by phytoplankton[3] and released to the atmosphere above the oceans. Phytoplankton is the collective name for the many types of microscopic plants, coming in a variety of shapes that dwell just below the ocean surface. Their name is derived from the Greek *phyton* ("plant") and *plagty* ("drifter"), and they are sometimes referred to as the "grasses of the sea." They are similar to land-based plants, containing chlorophyll and using sunlight for photosynthesis, which is why they are found close to the surface. Their prevalence is revealed by satellite images such as the one shown in Fig. 2.6 from NASA's terra satellite. Huge turquoise-colored regions show the presence of blooming phytoplankton. The DMS released by the plankton emerges from the sea and oxidises in the atmosphere. The resulting compounds condense into sulfur-containing nanoparticles that are carried high into the atmosphere.

A recent study of marine aerosol, generated above the North Atlantic and arriving at Mace Head on the west coast of Ireland, quantified the organic (life-produced) and inorganic contributions to the marine aerosol in different seasons [12]. A marked difference was found between periods of high biological activity (plankton blooms) in the summer and periods of low biological activity in the winter. Figure 2.7a shows the level of chlorophyll across the Atlantic in summer and winter, with (appropriately) green representing high chlorophyll levels. Notice how the summer bloom lights up the whole North Atlantic—an event of enormous scale. The bar graphs in Fig. 2.7b show the composition of the marine aerosol during the two periods separated into different types of particle in different size ranges. The first bar covering the size range 0.06–0.125 μm (60–125 nm) shows the nanoparticle abundance. In winter these are undetectable when measured as a mass fraction (though they would still dominate if measured as a number density), and the entire diagram is dominated by sea-salt particles.

---

[3]More precisely, it is released when plankton of the animal kind (zooplankton) eat plankton of the plant kind (phytoplankton).

Fig. 2.6   Phytoplankton bloom in the North Sea. Clouds of phytoplankton are the turquoise colored patches in this image acquired on June 27th 2003 by the MODIS instrument on NASA's Terra satellite. The land mass at the top right is Norway and Denmark is on the bottom right. Phytoplankton grow in nutrient-rich waters, and multiply very quickly; blooms big enough to be seen from space, like this one, can take only days to appear. Also visible are a number of streaky airplane contrails. Image reproduced courtesy of NASA (http://visibleearth.nasa.gov).

During the summer periods, water-soluble and water-insoluble organic nanoparticles generated by phytoplankton are prolific and comprise most of the particles.

The fact that life generates aerosol produces an important feedback mechanism in the climate. As described above, an increase in the density of atmospheric aerosol generates more cloud, which reduces the amount of sunlight reaching the sea surface and thus reduces the energy available for phytoplankton. If the phytoplankton are less prolific, the nanoparticle production rate via the DMS route is scaled back. In the context of global warming, this is a stabilizing effect because warmer seas encourage phytoplankton growth, which increases the rate of DMS production and thus the amount of cloud, which produces a cooling.

Not just the products of life but life itself is thrown out of the sea by the bursting bubble route that produces sea-salt particles. Among the soup of microscopic organisms that live near the ocean surfaces are viruses, which are themselves small enough to be part of the nanoworld. The ones thrown out of the sea join the general atmospheric aerosol of nanoparticles and act as CCNs. In the arctic they are thought to be a significant contribution to the CCNs responsible for clouds [13].

2.5   NANOPARTICLES IN SPACE

Finally in this chapter we will extend the discussion out beyond the Earth's atmosphere into space. To begin with, it is known that the constant stream of

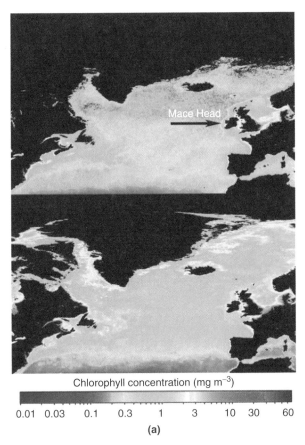

Chlorophyll concentration (mg m^{-3})

0.01 0.03   0.1   0.3   1   3   10 30   60

(a)

Fig. 2.7 Composition of particles produced by phytoplankton. (a) Seasonal average over 5 years of sea-surface chlorophyll concentrations in winter (top image) and spring (bottom image) obtained by the Sea-viewing Wide Field-of-view Sensor (SeaWiFS) instrument in low earth orbit (courtesy of SeaWiFS Project, NASA/Goddard Space Flight Centre and ORBIMAGE – see http://disc.gsfc.nasa.gov/oceancolor/scifocus/oceanColor/nab.shtml). The seasonal difference in biological activity is clear. The location of Mace Head where the aerosol composition was measured in shown in the top image. (b) Composition of aerosol in different size ranges. The region from 0.06 μm – 0.125 μm (60–125 nm) shows the nanoparticle abundance. In winter they are undetectable but during phytoplankton blooms they are abundant. The data is given for the different particle types: Sea salt (produced by the bubble bursting mechanism), NH4, non-sea-salt (nss) SO4, NO3, water-soluble organic carbon (WSOC) and water-insoluble organic carbon (WIOC). Reproduced with the permission of the Nature Publishing Group from C. O'Dowd et al. [12].

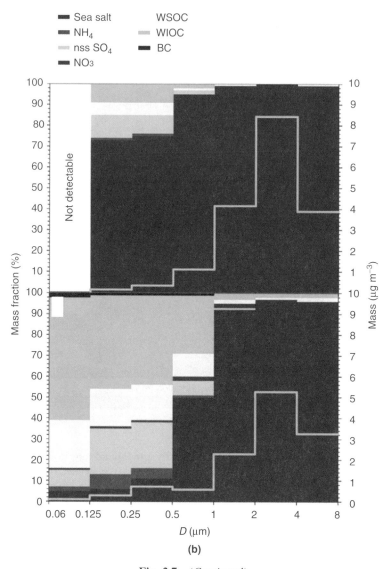

Fig. 2.7 (*Continued*)

fast charged particles, mostly protons that emanate from the sun, affect our climate. The details of the interaction are complex, and there may be several different mechanisms but a prominent one involving nanoparticles is that the cosmic rays entering the atmosphere leave a trail of ionized gas molecules that can act as nucleation centers for CCNs. Put simply, the cosmic rays encourage cloud formation. As described above, this can in turn affect the atmospheric aerosol load via, for example, the feedback mechanism involving DMS

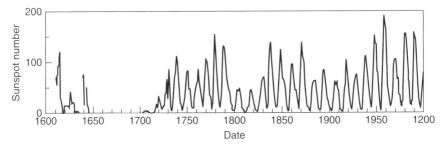

Fig. 2.8 Maunder minimum and the Mini Ice Age. Plot, using historical records, of the observed sunspot number over the last 400 years. The Maunder minimum between 1645 and 1715 coincides with the coldest part of the so-called Mini Ice Age in the 17th and 18th centuries in which unusually cold winters occurred. Reproduced from http://en.wikipedia.org/wiki/Maunder_minimum.

and phytoplankton. The relatively small flux of cosmic rays incident on the Earth can thus have a disproportionately large effect on atmospheric conditions. The amplifying effect arises from cosmic rays influencing the amount of atmospheric aerosol, mostly in the form of nanoparticles. The solar cosmic ray flux shows large variations with conditions of the sun—for example, the 11-year sunspot cycle. A possible illustration of how influential solar activity can be is the correlation between the period of the Maunder minimum—when very few sunspots were observed, indicating low activity and cosmic ray flux—and the coldest part of the so-called Mini Ice Age in the 17th and 18th centuries (Fig. 2.8).

Strong evidence for a direct link between atmospheric aerosol and solar activity comes from the Greenland Ice Sheet Project 2 (GISP2), which examines the depth profile of impurities in Greenland ice. This is a convenient way to determine atmospheric conditions in the past because any particular concentrations of chemicals or particles are frozen into the ice at a depth that depends on how long ago they were around. Analysis of the variations in particle concentration of the top 120 m, corresponding to the last 400 years, shows a correlation of the aerosol load with the sunspot number—that is, with the solar cosmic ray flux [14]. Not all cosmic rays come from the sun and there is a significant flux, especially of higher-energy particles, from sources outside our own galaxy. It is sobering to realize that events in the far reaches of the universe can have an influence on our climate.

It is clear that cosmic rays influence the density of atmospheric nanoparticles, but nanoparticles themselves do not stop at the top of the Earth's atmosphere and space itself is permeated by cosmic dust arising from a number of sources. Supernovae (Fig. 2.1e) have already been mentioned; but others include out-flowing material from carbon-rich stars, which is rich in silicon carbide and titanium carbide particles [15] as well as various forms of pure carbon particles including fullerenes (see Chapter 3). As with the particle populations measured in the Earth's atmosphere, when measured as the number density or surface area, it is nanoparticles (<100 nm) that dominate the distribution (Fig. 2.2). Thus nanoparticles provide a significant proportion of the solid surface area in space on which chemical reactions can take place.

Interstellar dust particles also accelerate the process of condensation of gas clouds by gravity to form stars and planets; thus nanoparticles were an important ingredient in the initial formation of our own sun and its planets including the Earth. It is interesting to note that the special properties of nanoparticles compared to bulk matter discussed in Chapter 1 are also important in this context. For example, a significant fraction of particles produced by supernova explosions contain iron (from the core of the exploding star) and are magnetic. The magnetic interaction between the particles in space, which is orders of magnitude stronger than their gravitational attraction, can enormously accelerate the process of condensation and for this to work the particles must be single domains—that is, permanently magnetized. As discussed in the previous chapter, this requires that they be smaller than a critical size of about 100 nm. Once stars and planets are formed, they produce interplanetary particles by various processes. For example, in our own solar system the Jovian satellite Io, which is has a very high volcanic activity, sprays vast quantities of particles into the rest of the solar system [16].

This chapter has by far the widest scope in this book, and each topic introduced here could easily occupy a book of its own. The treatment therefore has necessarily been superficial, but a number of references are given for a more in-depth study of various topics. The aim has been to give a flavor of the huge importance of naturally occurring nanoparticles in shaping our environment. In the next chapter we will bring our attention back into the research laboratory and discuss the fascinating world of carbon nanoparticles.

PROBLEMS

1. In a volcanic eruption, most of the mass of volcanic ash is distributed in particles with sizes in the range 1 μm to 1 mm, and the plume reaches a height of up to 20 km. Assume a prevailing wind with an average speed of 10 ms^{-1}. Use the equation in Advanced Reading Box 2.1 to calculate the maximum distance downwind of the volcano at which 1-mm-diameter particles and 1-μm-diameter particles are deposited. What size of particle can be expected to be deposited over the entire globe?

2. The table and graph below show the aerosol concentration in mg/m^3 as a function of particle diameter measured in a typical urban environment. Assuming that the average density of the material in the particles is 2000 kg/m^3, convert the data to show the number of particles per cubic meter as a function of particle diameter.

Particle Diameter (μm)	Mass per Unit Volume (mg/m^3)	Particle Diameter (μm)	Mass per Unit Volume (mg/m^3)
0.01	1.00E-05	0.8	0.069
0.02	8.00E-04	0.9	0.069
0.03	0.005	1	0.068
0.04	0.007	2	0.06
0.05	0.008	3	0.05
0.06	0.01	4	0.065
0.07	0.011	5	0.078
0.08	0.012	6	0.086
0.09	0.014	7	0.091
0.1	0.016	8	0.095
0.2	0.029	9	0.098
0.3	0.037	10	0.1
0.4	0.047	20	0.091
0.5	0.059	30	0.072
0.6	0.065	40	0.055
0.7	0.068	50	0.04

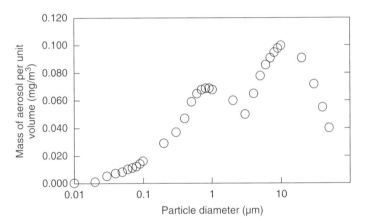

3. Derive equation (2.4) in Advanced Reading Box 2.2 by considering the change in the energy of the surface tension of a liquid drop due to the shrinkage resulting from the evaporation of a single molecule.

4. Describe how phytoplankton produce an important feedback mechanism that helps to reduce global warming.

REFERENCES

1. W. C. Hinds *Aerosol Technology*, John Wiley & Sons, New York, 1999.

2. J. J. Seinfeld and S. N. Pandis *Atmospheric Chemistry and Physics: From Air Pollution to Climate Change*, John Wiley, & Sons, New York 1997.

3. J. Roberts and J. H. Quastel, Particle uptake by polymorphonuclear leucocytes and Ehrlich ascites-carcinoma cells, *Biochem. J.* **89** (1963), 150–156.

4. P. M. Hoet, I. Brüske-Hohlfield, and O. V. Salata, Nanoparticles—known and unknown health risks, *J. Nanobiotechnol.* **2** (2004). Available at http://www.jnanobiotechnology.com/content/2/1/12.

5. L. C. Renwick, K. Donaldson, and A. Clouter, Impairment of alveolar macrophage phagocytosis by ultrafine particles, *Toxicol. Appl. Pharmacol.* **172** (2001), 119–127.

6. K. Donaldson, L. Tran, L. Albert. Jimenez, R. Duffin, D. E. Newby, N. Mills, W. MacNee, and V. Stone, Combustion-derived nanoparticles: A review of their toxicology following inhalation exposure, *Particle Fibre Toxicol.* **2** (2005). Available at http://www.particleandfibretoxicology.com/content/2/1/10.

7. M. Kreilgaard, Influence of microemulsions on cutaneous drug delivery, *Adv. Drug Delivery Rev.* **54** (2002), S77–S98.

8. N. Hussain, V. Jaitley and A. T. Florence, Recent advances in the understanding of uptake of microparticulates across the gastro-intestinal lymphatics, *Adv. Drug Delivery Rev.* **50** (2001), 107–142.

9. Health and Safety Executive Health effects of nanoparticles produced for nanotechnologies, HSE Hazard assessment document EH75/6, December 2004. Available at http://www.hse.gov.uk/horizons/nanotech/healtheffects.pdf.

10. A. J. Walton, *Three Phases of Matter*, Clarendon, Oxford, 1983, Chapter 13, pp. 456–460.

11. A. Clarke, V. Kapustin, S. Howell, K. Moore, B. Lienert, B, S. Masonis, T. Anderson, and D. Covert, Sea-salt size distributions from breaking waves, *J. Atmos. Oceanic Technol.* **20** (2003), 1362–1374.

12. C. O'Dowd, M. C. Facchini, F. Cavalli, D. Ceburnis, M. Mircea, S. Decesari, S. Fuzzi, Y.-J. Yoon, and J.-P. Putaud, Biogenically driven organic contribution to marine aerosol, *Nature* **431** (2004), 676–680.

13. M. Higgins, Arctic clouds: They are alive!, December 2005, http://cires.colorado.edu/~higginsm/clouds/index.html.

14. J. Donarummo, Jr., M. Ram, and M. R. Stolz, Sun/dust correlations and volcanic interference, *Geophys. Res. Lett.* **29** (2002), 75-1–75-4.

15. D. Clement, H. Mutschke, R. Klein, and T. Henning, New laboratory spectra of isolated SiC nanoparticles: Comparison with spectra taken by the Infrared Space Observatory, *Astrophys. J.* **594** (2003), 642–650.

16. F. Postberg, S. Kempf, R. Srama, S. F. Green, J. K. Hillier, N. McBride, and E. Grün, Composition of Jovian dust stream particles, *Icarus* **183** (2006), 122–134.

CHAPTER 3

Carbon Nanostructures: Bucky Balls and Nanotubes

This chapter focuses on nanostructures produced by carbon. It may seem strange to devote a chapter to a single element, but there is such a plethora of nanostructures composed of carbon that it is easy to justify several books (as indeed there are), never mind a chapter on them. Here we will pass lightly over the subject and discuss how and why these structures form and their basic properties. Their use in nanotechnology is described in Chapter 5, but their applications are diverse and references to carbon nanostructures can be found throughout the rest of this book.

3.1 WHY CARBON?

Carbon is a light, simple element: Its atoms contain just six electrons, two of them being core ($1s$) electrons and the remaining four ($2sp$) available for bonding with other atoms. It is the slightly schizophrenic nature of the chemical bonding by these four that naturally gives carbon a diversity of forms. The details of the environment in which the atoms come together (pressure, temperature, etc.) determine the types of bonds. For example, we are very familiar with the two very different bulk forms (allotropes) of carbon—that is, graphite and diamond that result from different types of bonds. The bonding in diamond and graphite and why they adopt their particular crystal structures is illustrated in Fig. 3.1.

The diagrams on the left show the charge distribution associated with the four bonding electrons for the two crystal structures. It is important to realize that these have no meaning out of the context of bonding to other atoms (see Advanced Reading Box 3.1). They are the charge distributions that would be found if we suddenly plucked a carbon atom out of its crystal and somehow kept the electronic charge distribution associated with the atom frozen. In the case of diamond, the four bonding electrons produce a tetrahedral charge distribution around each atom and so the atoms come together along these mutual bonds forming a tetrahedral arrangement. The bonds (covalent σ bonds) are strong,

Introduction to Nanoscience and Nanotechnology, by Chris Binns
Copyright © 2010 John Wiley & Sons, Inc.

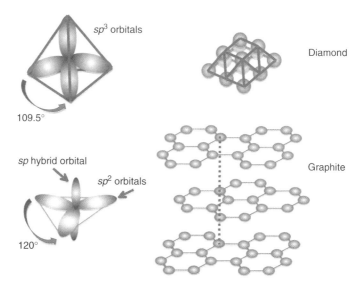

Fig. 3.1   Carbon bonding in diamond and graphite. Illustration of bonding of carbon in diamond and graphite crystal structures. There are four electrons in each carbon atom that produce bonding with nearest neighbors and in diamond (top) the charge distribution associated with these electrons forms a tetrahedral structure (sp^3 orbitals) around each atom. The atoms thus come together in a tetrahedral arrangement. In graphite (bottom) the bonding electrons form a charge distribution of three equally spaced lobes in a plane (sp^3 orbitals) with the charge distribution of the fourth being out of the plane (sp hybrid orbital). In graphite the carbon atoms are thus strongly bonded in a hexagonal arrangement in sheets with weak bonding between the sheets (see Advanced Reading Box 3.1).

giving diamond its extreme hardness. Since all the electrons are involved with bonding, they don't interact with light, thereby making diamond a good insulator and transparent though impurities and defects can give it an intrinsic color.

ADVANCED READING BOX 3.1—ORBITAL HYBRIDIZATION IN CARBON BONDING

The ground-state electronic configuration of a free carbon atom is $1s^2 2s^2 2p_x^1 2p_y^1$; however, since the $2p_x, 2p_y$, and $2p_z$ orbitals are degenerate, it is not necessary to distinguish them. The $1s$ electrons with a binding energy of around 250 eV are core levels and not involved in bonding. Since the 2s shell is full, at first glance one might expect carbon to form two bonds with the unfilled $2p$ orbitals. In fact the interaction with other atoms mixes (hybridizes) the $2s$ and $2p$ electrons to form four bonding orbitals with two different types of mixing. With a given set of conditions, interaction with hydrogen or other carbon atoms produces four equivalent sp^3 orbitals (top portion of Fig. 3.1) whose charge distribution is four lobes in a tetrahedral arrangement. Such sp^3 orbitals form tetrahedral molecules—for example, CH_4,—or the tetrahedral diamond structure. In other environments it is possible for the $2s$ and $2p$ orbitals to hybridize into three sp^2 orbitals whose charge distribution forms three equally spaced lobes in a plane plus a remaining p orbital. It is this configuration that forms the planar graphite structure (bottom portion of Fig. 3.1). Both the sp^3 and sp^2 orbitals form strong covalent σ bonds; and in the case of diamond, all atoms are locked into a tetrahedral arrangement by the tetrahedral network of σ bonds giving diamond its hardness. In the case of graphite, there is a strong network of σ bonds in each plane (stronger than diamond in fact) but weak van der Waals bonding (see Advanced Reading Box 3.4) between planes so that they slip relatively easily with respect to each other, giving graphite its apparent softness and lubricant quality. The electrons in diamond are all involved with bonding and are tightly bound with a high binding energy so that diamond is a very wide bandgap insulator and does not interact with visible light. It is thus transparent though defects and impurities can color it. In graphite the three sp^2 orbitals are involved in bonding; but the remaining electron is weakly bound and forms a metallic band structure, making graphite an electrical conductor.

In the case of graphite the charge associated with the bonding electrons forms three equally spaced lobes in a plane with the remaining bonding electron charge distributed out of the plane. In this configuration the bonding electrons form a strong hexagonal network of bonds with other carbon atoms where the in-plane bonding is even stronger then diamond. The planes, however, interact weakly (by van der Waals bonding—see Advanced Reading Box 3.4) and can slip relatively easily with respect to each other. This gives graphite its apparent softness and lubricant quality. The remaining electrons whose charge is distributed out of the plane form a band of free electrons as in a metal, making graphite an electrical conductor.

It is the graphite structure that is most relevant to this chapter. A single atomic plane of graphite, known as *graphene* (Fig. 3.2), is a flat hexagonal arrangement of carbon atoms with very strong bonding in-plane. Because it has become possible to isolate and study these individual atomic planes (Fig. 3.2), graphene has recently become the focus of a significant research effort. The flat perfect hexagonal arrangement, however, is hard to isolate and quite unstable. The existence of defects such as pentagonal or heptagonal rings will bend and warp the sheets, and they have a tendency to roll up to form tubes or spherical cages. This is the basis of the formation of fullerenes, including bucky balls and nanotubes.

3.2   DISCOVERY OF THE FIRST FULLERENE: C_{60}

The first discovery of a fullerene occurred at Rice University during a feverish 10 days of activity in September 1985. The team included Harry Kroto (then at

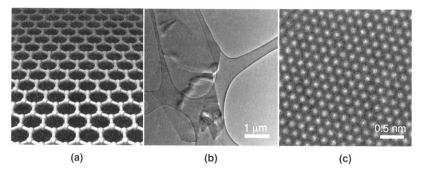

(a)   (b)   (c)

Fig. 3.2 Graphene sheet. (a) A single hexagonal layer of carbon atoms (graphene) is the starting point for the formation of fullerenes. Only defect-free hexagonal arrangements will remain flat. Any defects such as pentagonal or heptagonal rings will curl the sheet and it is unstable against rolling up to form tubes or shell-like structures such as nanotubes and bucky balls. Reproduced from Wikipedia web page: (http://en.wikipedia.org/wiki/Graphene). (b) Transmission Electron Microscope (TEM) image of a sheet of carbon only a few atoms thick (rumbled layer) on a lacy carbon support (see Chapter 4, Section 4.4.6). Towards the edges this thins to a single graphene layer. (c) TEM image of a single graphene layer showing the hexagonal arrangement of carbon atoms. Reproduced with the permission of the Nature Publishing Group from J. H. Warner et al. [1].

Sussex University, UK), Bob Curl, and Richard Smalley, who all shared the 1996 Nobel Prize for chemistry in recognition of their discovery. The text of the Nobel lecture by Harry Kroto that related the dramatic story was published in *Reviews of Modern Physics* [2]. At the time, Kroto was mainly interested in the spectra of carbon chain molecules found in interstellar space and their synthesis in the laboratory. During a visit to Rice University in 1985, he was shown an apparatus for generating beams of metal nanoparticles based on laser ablation (see Chapter 4, Section 4.1.2). He suggested that the environment inside the machine might replicate some of the conditions found around certain types of stars and that installing a carbon target might replicate carbon chain molecules found around those stars. The suggestion was taken up; and the mass spectrum of species produced by the machine, as measured by a time-of-flight mass spectrometer, obtained on September 4, 1985, is shown in Fig. 3.3. The dominant peak at a time of flight of about 47 μs corresponds to a mass of 720 atomic mass units (amu), that is, 60 carbon atoms. By varying the condition in the machine, it was found to be possible to produce mass spectra with virtually nothing but C_{60} molecules. Somehow this molecule is exceptionally stable and under the right conditions is formed in preference to all others.

The rest is history, and it is now known that C_{60} is the cage-like molecule shown in Fig. 3.4. In fact, this structure for a 60-atom carbon molecule had already been predicted years earlier (1970) in an article in Japanese by Eiji Osawa [3], and even earlier (1966) David Jones [4] had surmised that introducing pentagonal rings into a graphene sheet would cause it to curl up into a hollow balloon. Because of the resemblance of cage-like molecules such as the one in Fig. 3.4 to

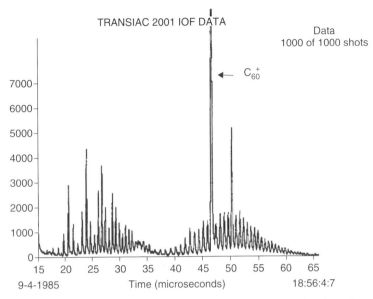

Fig. 3.3   First detection of C_{60}. Time of flight mass spectrum of carbon clusters produced in a laser ablation source (see Chapter 4, Section 4.1.2) taken on 4th September 1985 showing dominant peak due to C_{60}. Reprinted with permission from H. Kroto [2]. Copyright 1997 by the American Physical Society.

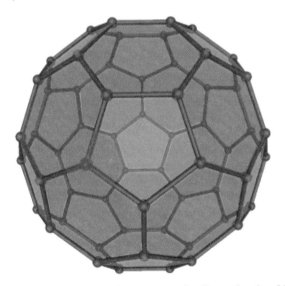

Fig. 3.4   Structure of C_{60}. The atomic structure of a C_{60} molecule of bucky ball, first suggested by Osawa [3] in 1970 and then synthesized in 1985 at Rice University [2]. The closed shape can be perfectly tiled by 12 pentagonal and 20 hexagonal tiles. It is the smallest cage that can be built out of pentagons and hexagons without having adjacent pentagons (see text).

geodesic domes designed by the architect Richard Buckminster Fuller, the term buckminsterfullerene has been coined for such structures, or, more affectionately, *bucky balls*. The general class of carbon molecules including graphene and all the tube and balloon structures it forms on curling up are known as *fullerenes*.

3.3   STRUCTURAL SYMMETRY OF THE CLOSED FULLERENES

Pentagonal rings are the key to understanding the structural stability of C_{60} and other closed fullerenes. Examining the molecule closely, we see that it has 20 hexagonal and 12 pentagonal faces. This is a particular member of a general class of polyhedral shapes constructed out of polygons that have engrossed mathematicians since ancient times. The fascination is in forming closed shapes out of polygonal tiles and finding rules for the formation of perfect closure with no gaps or overlap of tiles. In 1620, Descartes noticed a general rule connecting the number of faces, edges, and corners in a closed polyhedron that was later proved by Euler in 1752 and has since been known as Euler's theorem. If we manage to construct a perfect closed shape with polygonal tiles, then if F denotes the number of faces or tiles, E the number of edges, and V the number of vertices or corners, Euler proved that the formula

$$V - E + F = 2$$

must be satisfied. Figure 3.5 shows a few examples that verify the formula for a cube, a tetrahedron, and a C_{60} molecule.

Euler's formula can be used to determine whether it is possible to form a closed shape out of a given set of tiles; for example, it is easy to prove that it is impossible to form a closed shape using only hexagons. If one imagines starting to build the three-dimensional structure with the first three hexagonal tiles brought together along their edges, it is easy to verify that each pair shares an edge ($E = F/2$) and that each triple shares two vertices ($V = 3F/2$) so that $V - E + F = 0$, proving that completing a closed shape without gaps is impossible unless one also uses tiles with a different shape. In fact, it was realized by the ancient Greeks that there are only three polygons that can form closed shapes on their own—that is, a square a triangle and a pentagon.

In the case of C_{60} we need to mix pentagonal faces with the hexagonal ones, and here again the power of Euler's theorem comes to the fore. It can be proved that there are exactly 12 pentagons in the closed cage, irrespective of the number of hexagons and carbon atoms; using more hexagons simply makes the cage larger (see Advanced Reading Box 3.2). Remembering that pentagonal rings are defects, as far as carbon bonding is concerned, it is clear that having adjacent pentagons will introduce a great deal of strain and decrease the stability of the molecule. If we then inquire as to which is the smallest fullerene (with 12 pentagonal rings) that can be formed by having no adjacent pentagonal rings, the answer is the C_{60} structure in Fig. 3.4. This elegant combination of chemistry and structural symmetry dictates that the molecule must take this structure.

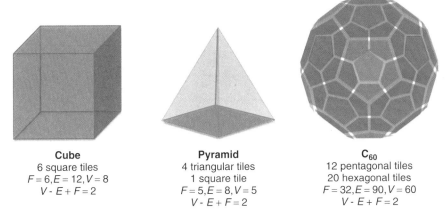

Cube	**Pyramid**	**C_{60}**
6 square tiles	4 triangular tiles	12 pentagonal tiles
$F = 6, E = 12, V = 8$	1 square tile	20 hexagonal tiles
$V - E + F = 2$	$F = 5, E = 8, V = 5$	$F = 32, E = 90, V = 60$
	$V - E + F = 2$	$V - E + F = 2$

Fig. 3.5   Euler's theorem. Verification of Euler's theorem for a cube, a pyramid and a C_{60} molecule.

ADVANCED READING BOX 3.2—USING EULER'S THEOREM TO PROVE A FULLERENE CAGE CONTAINS 12 PENTAGONS

Assume that fullerene is composed of p pentagons and h hexagons; then the number of faces is

$$F = p + h.$$

Each pair of adjoining tiles shares a single edge, irrespective of whether the adjoining tiles are hexagons or pentagons, so the number of edges in the polyhedron must be half the number of edges presented by the isolated tiles; that is,

$$E = \frac{5p + 6h}{2}.$$

Every three edges from the isolated tile set (i.e., irrespective of whether they are shared) defines a single vertex, so

$$V = \frac{5p + 6h}{3}.$$

Thus applying Euler's theorem gives

$$\frac{5p + 6h}{3} - \frac{5p + 6h}{2} + p + h = 2$$

that is,

$$\frac{5}{3}p + 2h - \frac{5}{2}p - 3h + p + h = 2$$

> or
>
> $$p = 12.$$
>
> Thus the number of hexagons doesn't matter, but there must be 12 pentagons if the tiling is to form a closed cage.

We can thus verify that the finished product is particularly stable if it is the well-known soccer ball structure. It is reasonable to ask, however, exactly how, in the hellish $5000°C$ environment of the laser ablation source where carbon atoms are bonding and re-evaporating, the C_{60} structure actually forms. More than 20 years after the discovery of C_{60}, this is still not completely resolved, but the general consensus is that, initially, sections of warped graphene sheet will close into giant fullerenes with more than 1000 carbon atoms and then, in the high-temperature environment, evaporate carbon atoms and shrink down to the smaller fullerenes. This process has been captured recently in a beautiful series of transmission electron microscope (TEM; see Chapter 4, Section 4.4.6) images [5] and a video available online [6]. In the experiment the starting structure was a multi-wall carbon nanotube (see Section 3.8), which was heated to more than $2000°C$ by passing a current through it. As observed in Fig. 3.6a the innermost wall of the tube collapses to form two giant fullerenes containing about 1300 and 1100 atoms. The sequence of images at later times (Figs. 3.6b–3.6e) shows both fullerenes shrinking by evaporating carbon atoms, with the initially smaller one bobbing in and out of the picture. From Fig. 3.6f to Fig. 3.6n the remaining fullerene continues to shrink right down to the size of C_{60} (Fig. 3.6n. In the following two images it appears to open and then completely vanish. As we shall discuss below (Section 3.4), it is possible to get fullerenes smaller than C_{60}; but they are not as stable, and below a critical size the strain will cause the fullerene to "explode" into fragments.

3.4 SMALLER FULLERENES AND "SHRINK-WRAPPING" ATOMS

The process described above whereby a fullerene shrinks by evaporating carbon atoms is often referred to as *shrink-wrapping*. In the years following 1985, when there was still a debate as to whether C_{60} really was a hollow cage structure, some elegant experiments demonstrated shrink-wrapping of fullerenes by exposing C_{60} molecules to high-power UV laser radiation in order to evaporate atoms from the cage. It was found, using mass spectrometry [7], that fullerenes containing $32–80$ atoms shrank by evaporating C_2 carbon atom pairs, not triplets or individual carbon atoms. This in itself is strong evidence for the proposed hollow cage structure because only by evaporating pairs can the next smaller, even product produce a closed cage. Small pieces of graphene sheet would be much more likely to evaporate the more stable C_3 triplets. Interestingly, fragments with

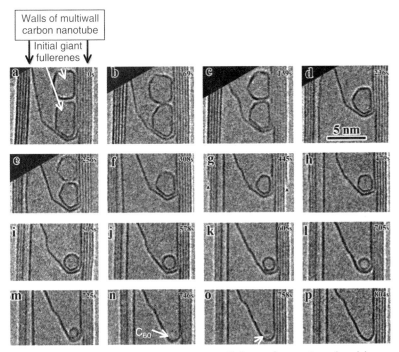

Fullerene fragments and vanishes

Fig. 3.6   Direct imaging of the formation of C_{60} from giant fullerenes. Series of TEM images recorded as a function of time of the formation of fullerenes inside a multi-walled carbon nanotube upon heating to over 2,000°C by passing a current through it. The initial stage shown in (a) is the formation of two giant fullerenes. At later times both fullerenes shrink by evaporation of carbon atoms with one disappearing out of the frame after (e). The remaining fullerene shrinks down to a single C_{60} molecule at (n). Finally in the last two frames the fullerene is seen to fragment and disappear. Reprinted with permission from J. Y. Huang et al. [5]. Copyright 2007 by the American Physical Society.

less than 32 atoms always did evaporate such triplets with the implication that C_{32} is the smallest closed-cage fullerene that is stable. As the molecule shrinks from C_{60}, there are increasing numbers of adjoining pentagonal rings, and the strain resulting from too many of these causes the cage to burst open. This was observed directly in the final two TEM images shown in Fig. 3.6. Although C_{32} is the smallest stable closed-cage fullerene observed when shrinking from larger sizes, even smaller closed-cage fullerenes are observed in mass spectra in cluster sources, showing that it is possible to build them bottom-up. A particularly prominent peak at C_{28} was found in early mass spectra, which has been ascribed to the structure shown in Fig. 3.7 [2]. This contains the usual 12 pentagonal faces and just two hexagonal faces. This violates the no abutting pentagon rule, but it has a particularly favorable electronic structure. Finally, semi-stable fullerenes

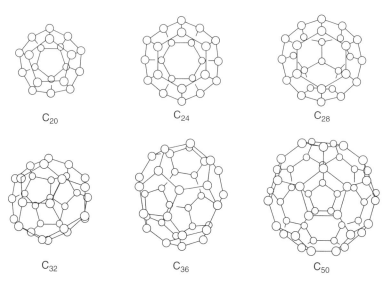

Fig. 3.7 Small fullerene structures. Structure of closed-cage fullerene clusters containing less than 60 atoms. They all violate the 'no-abutting pentagons' rule but they are able to form under suitable conditions. Shrink-wrapping experiments observe the fullerenes down to C_{32}, below which only fragments are observed. Smaller fullerenes have been seen in mass spectra from cluster sources showing they can be built bottom-up. The smallest possible structure is C_{20} containing only the 12 pentagonal faces required for closure. Reprinted with permission from H. Kroto [2]. Copyright 1997 by the American Physical Society.

can be predicted all the way down to the C_{20} structure also shown in Fig. 3.7, which contains only the 12 pentagonal faces required for closure.

A particularly innovative form of the shrink-wrapping experiments was to introduce metal atoms into the vapor in which the fullerenes formed by impregnating the carbon target with the metal of interest. This was done with lanthanum soon after C_{60} itself was discovered, and compelling evidence was presented that the lanthanum was trapped inside the carbon cage [8]. Such a carbon–metal complex is termed an endohedral fullerene (from the Greek "endo" meaning within), and the structure is shown schematically in Fig. 3.8 for a metal atom within a C_{60} molecule. The discovery of endohedral fullerenes led to a new series of laser evaporation experiments in which the surrounding carbon cage was shrink-wrapped around the central atoms. One such study demonstrated that, while the smallest pure fullerene with a closed cage contains 32 atoms, the smallest closed-cage fullerene containing a potassium atom is C_{44} and the smallest one containing a cesium atom is C_{48}. These observations support the simple explanation that the cage shrink-wraps the atom and that the smallest cage that can be formed depends on the metal atom diameter. Cesium is a slightly larger atom than potassium, so the minimum size cage contains four more carbon atoms than in the case of an

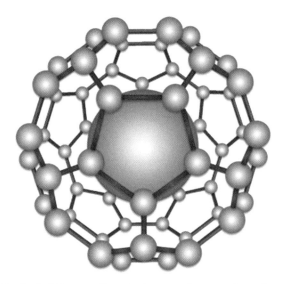

Fig. 3.8   Endohedral fullerene. Introducing metal atoms into the vapor in which fullerenes are created produces metal atoms trapped inside the fullerene cage [8].

enclosed potassium atom. Both of these are larger than the minimum size empty cage of 32 carbon atoms.

Since the early experiments, a number of different atoms and molecules have been placed inside fullerenes. Many can be formed by introducing a metal vapor impurity into the carbon vapor in which the fullerenes are forming. Later experiments showed that it is also possible to implant ionized metal atoms into an already formed empty cage. The fullerenes formed by processes such as laser ablation are commercially available as powders. These can then be heated in vacuum in a crucible, and the fullerenes can be evaporated and deposited onto a surface as if they were artificial atoms as shown in Fig. 3.9 (see Fig. 3.12 and Fig. 1.1, Chapter 1 for images of deposited C_{60} fullerene films). In fact, since the bonding between fullerenes is quite weak, they can be made to evaporate as complete fullerenes at a relatively modest temperature of about $500°C$. If the metal ions are fired at the same surface at the right energy (~50 eV), the ones that impact an empty fullerene will break through the wall, which then heals itself enclosing the metal atom. The energy imparted to the fullerene by the process will eject it from the surface [9], allowing it to be collected. Endohedral fullerenes containing various rare-earth atoms (terbium, thulium, neodymium, europium, and erbium) are now commercially available.

3.5   LARGER FULLERENES

C_{60} is the most stable molecule in the fullerene series; but as stated above and proved, using Euler's theorem in Advanced Reading Box 3.2, any closed shell

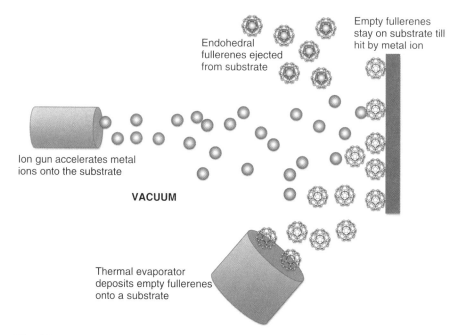

Empty fullerenes stay on substrate till hit by metal ion

Endohedral fullerenes ejected from substrate

Ion gun accelerates metal ions onto the substrate

VACUUM

Thermal evaporator deposits empty fullerenes onto a substrate

Fig. 3.9   Endohedral fullerenes produced by implanting ions. Fullerenes can be evaporated as complete molecules at relatively modest temperature from a crucible in vacuum onto a substrate to form fullerene films (for example See Fig. 3.12 and Fig. 1.1, Chapter 1). If ions from an ion gun are fired at the same substrate at the right energy (~50 eV) the ones that impact a fullerene will be implanted into the cage, which will then heal itself around the atom. The energy imparted to the fullerene will eject it from the surface [9] so empty fullerenes will stay on the surface while there will be a steady stream of endohedral fullerenes coming off the surface.

will be produced by 12 pentagons, irrespective of the number of hexagons. Thus there should be a whole series of viable structures formed by increasing the number of hexagons, which for n atoms in the fullerene is the number ($n/2 -$ 10). Molecular stability depends not only on particularly symmetric geometric structures but also on the stability of the resulting electronic shells. It is found that there is a series of stable fullerenes containing $60 + (k \times 6)$ atoms, where the k represents the integers: $0,2,3,4,5......$ (note that $k = 1$ is not included in this series). As the fullerene size increases, the average distance between the pentagons increases, and this continues all the way up to graphene (the "infinite fullerene"). Within this series, there are especially stable numbers of atoms—that is, special values of n when, for example, the closed shell has a low-energy morphology. These are known as magic numbers, with $k = 0$ (C_{60}) being the first magic number. Figure 3.10 shows the fullerene C_{540} ($k = 80$), which is also a magic number since the closed cage forms a perfect icosahedral (20-sided, from the Greek "ikosi" meaning 20) structure. Magic numbers are a general feature of

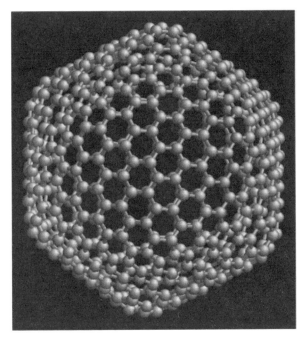

Fig. 3.10   Icosahedral structure of C_{540}. The fullerene C_{540}, which corresponds to a magic number ($k = 80$). It forms a perfect icosahedron (20-sided cage), which is especially stable. Reproduced from Wikipedia web page: http://en.wikipedia.org/wiki/Fullerene.

small nanoscale particles whether or not they form closed cages. For example, clusters of metal atoms also show magic numbers—that is, especially abundant clusters with a given number of atoms emerging from nanoparticle sources (see Chapter 4, Section 4.1.3).

The constant number of pentagons in the fullerene series has been confirmed by shrink-wrapping experiments such as the one shown in Fig. 3.6 [5]. It is found that the rate of evaporation of C_2 pairs is constant, irrespective of the size of the fullerene, which is odd since normally one would expect the rate to depend on the number of atoms in the molecule. As shown in Advanced Reading Box 3.3, at the temperatures normally used to evaporate carbon from fullerenes in shrink-wrapping experiments, the probability of removing a C_2 pair from a pentagonal ring is more than a million times higher than that from a hexagonal ring. So the evaporation rate will only depend on the number of pentagons, which according to Euler's theorem is a constant 12. When a C_2 pair is removed from a fullerene, this introduces new defects, from which carbon is also more easily evaporated. A complex process then ensues [5], but the end result is that the fullerene heals itself into one of the $60 + (k \times 6)$ cages, again with 12 pentagons so the evaporation rate remains constant through the shrink-wrapping process.

ADVANCED READING BOX 3.3—EVAPORATION RATE OF C_2 PAIRS FROM PENTAGONAL RINGS

Electronic structure calculations show that it takes 3 eV less energy to remove a C_2 atom pair from a pentagonal ring than a hexagonal ring. The typical temperature at which shrink-wrapping by evaporation experiments are carried out is $\sim 2000°C$ or ~ 2300 K. The ratio of probabilities of removing a C_2 pair from a pentagon instead of a hexagon is given by the ratio of the Boltzmann factors:

$$\frac{\exp(E_1/kT)}{\exp(E_2/kT)} = \exp((E_1 - E_2)/kT),$$

where E_1 is the energy required to remove a C_2 pair from a pentagonal ring and E_2 is the energy required for removal from a hexagonal ring. The difference $(E_1 - E_2) = 3$ eV $= 4.8 \times 10^{-19}$ J so the ratio in Boltzmann factors is

$$\exp(4.8 \times 10^{-19}/(1.38 \times 10^{-23} \times 2300)) = 3.7 \times 10^6.$$

The $60 + (n \times 6)$ series of fullerenes starting from C_{60}, described above, is only one of the possible series. It was observed, right from the first discovery of C_{60}, that C_{70} is also a highly stable molecule. Any mass spectrum showing the C_{60} peak always has C_{70} as well, as is evident in Fig. 3.3, where it shows up as the second most intense peak after C_{60}. It was soon realized that this structure, with 10 extra atoms, could be constructed by adding a belt of 5 extra hexagonal faces around the waist of the C_{60} molecule to produce the elliptical fullerene shown in Fig. 3.11.

Including elliptical and tubular fullerenes there are two extra series with stable numbers of atoms given by $70 + 30k$ $(k = 0, 1, 2, 3, \ldots)$ and $84 + 36k$ $(k = 0, 1, 2, 3, \ldots)$. Going from one size to the next involves a tubular extension about the long axis producing an increasing elongation. For large numbers these start to look like carbon nanotubes, described below, but in the case of nanotubes it makes more sense to consider the starting structure as a rolled-up graphene sheet.

3.6   ELECTRONIC PROPERTIES OF INDIVIDUAL FULLERENES

For technological applications the electronic properties of fullerenes are of great interest, but at present the only information available on the electronic states in free molecules has been obtained by theoretical calculations [10]. In order to study the electronic behavior experimentally (and indeed for applications), it is necessary to support the molecules on a substrate. Using preformed fullerene powder, it is straightforward to vacuum-evaporate complete fullerenes onto surfaces as in Fig. 3.9, and the growth of fullerene films (mostly C_{60}) has been

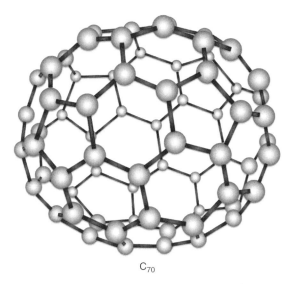

C_{70}

Fig. 3.11   Structure of C_{70}. The structure of C_{70} consisting of 25 hexagonal faces plus the 12 pentagonal faces required for closure. The elliptical molecule is the first in the series with numbers of atoms equal to $70 + 30k$ ($k = 0, 1, 2, 3, \ldots$) with each next larger member produced by a tubular extension about the long axis.

investigated on various substrates including Si, Ag, and Au. If the substrate is at room temperature or below, the molecules are not strongly bonded to it; and although they do not spontaneously diffuse around, they can be moved relatively easily by modest heating. They can also be deliberately moved in a controlled way using the probe tip of a scanning tunneling microscope to form controlled structures such as the now famous C_{60} abacus created at the IBM research laboratories in Zurich in 1996 (see Chapter 4, Section 4.4.2). This is a hint of the possibilities of radical nanotechnology, which attempts to build machines with molecular-sized components (see Chapter 7).

ADVANCED READING BOX 3.4—VAN DER WAALS FORCES

There are various interactions between atoms resulting from a redistribution of electronic charge—for example, ionic bonds or covalent bonds, such as those that produce the fullerene cages. Atoms or molecules in which the electronic configuration is particularly stable such as rare-gas atoms or C_{60} molecules do not form chemical bonds with each other, but in the absence of any bonds they still condense into solids. If no other interaction is present, electrically neutral quantum objects will still attract each other via the van der Waals interaction, named after the Dutch physicist Johannes van der Waals. Although the atoms are electrically neutral, the negative and positive charges are separated spatially;

and if their average distribution in space is disturbed, an electric dipole will result. The charge distributions in atoms or molecules are subject to quantum fluctuations so that at any instant they will present a dipole. In fact the charge fluctuations are related to zero-point quantum fluctuations of the vacuum electromagnetic field (see Chapter 8), but that discussion is not required here. The important point is that the instantaneous dipoles on neighboring atoms will tend to anti-correlate, and the opposing dipoles will attract each other. The effect is very weak compared to other types of interaction; and if covalent or ionic bonds are present, they will dominate the van der Waals force.

Although the effect is quantum mechanical, the strength of the attraction can be estimated classically. Suppose an atom or molecule has an instantaneous electrical polarization $p = qd$, where the dipole consists of charges $+q$ and $-q$ separated by a distance d. At a distance r from the dipole, which is sufficiently far that $r \gg d$, the electric field due to the dipole is given by

$$E(r) = \frac{p}{4\pi\varepsilon_0 r^3}.$$

A second atom in this field will polarize to produce a dipole, p, given by

$$p' = \alpha E(r) = \frac{\alpha p}{4\pi\varepsilon_0 r^3},$$

where α is the polarizability of the atom. The energy of interaction between the two atoms is thus

$$u(r) = -p'E(r) = -\alpha(E(r))^2,$$

that is,

$$u(r) \propto -\frac{1}{r^6},$$

which drops off rapidly with distance. For condensation this is only part of the story since at sufficiently small separations the atoms repel each other as the electron orbitals start to overlap. This is an even shorter-range interaction; that is, the inverse power of r must be larger than 6. For ease of further computations, this is usually taken to be 12; thus the full potential for atoms or molecules attracting each other only by van der Waals forces is given by

$$u(r) = -\frac{A}{r^6} + \frac{B}{r^{12}}.$$

This is the "Lennard-Jones potential." At their equilibrium separation the energy of attraction of two atoms via the van der Waals force is typically a few milli-electron-volts compared to covalent or ionic bonds, which are typically more than 1 eV—that is, more than a hundred times stronger.

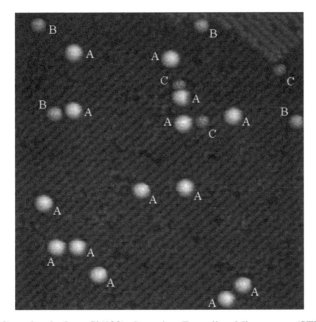

Fig. 3.12   C$_{60}$ adsorbed on Si(100). Scanning Tunneling Microscopy (STM) image of C$_{60}$ molecules adsorbed on a Si(100) surface. The light dots in the background are the Si atoms at the surface and the larger light blobs are individual C$_{60}$ molecules. The sample was prepared by initially depositing a sparse coverage of C$_{60}$ on the substrate at room temperature, heating to 600°C for a few seconds and then when the substrate had cooled depositing some more C$_{60}$. Thus fullerenes can be compared before and after annealing in the same image. Type A is found only in room temperature depositions and is bonded to the surface by Van der Waals forces (see Advanced Reading Box 3.4). Types B and C are obtained after annealing and their apparent smaller size is due to their sitting lower on the surface as a result of covalent bonds formed with the Si. Note the changes of internal features seen in the different types due to a redistribution of electronic states after bonding (see text). Reproduced with permission from Xiaowei Yao, Richard K. Workman, Dong Chen and Dror Sarid, Optical Sciences Centre, University of Arizona (see http://mcallister.com/papers/sarid.html).

The details of bonding to surfaces are clearly more complex as shown by the work of the Optical Sciences group at the University of Arizona, who obtained the scanning tunneling microscope image (shown in Fig. 3.12) of C$_{60}$ molecules deposited on an Si surface. The small white dots in the background are Si atoms, and the larger white blobs are the fullerenes. The sample was prepared in two stages; that is, a sparse coverage of C$_{60}$ was first deposited at room temperature and then the substrate was heated to 600°C for a few seconds. When it had cooled back to room temperature, some more C$_{60}$ molecules were deposited so fullerenes before and after annealing can be directly compared on the same image. Interestingly, although all the white blobs are C$_{60}$, they apparently come in three different sizes, labeled A, B, and C in the image. Before annealing, only type A can be observed; after annealing, only types B and C are found. The electrons in both the C_{60} and the Si surface are in stable configurations and without heating do not modify their structure to form chemical bonds between the adsorbed molecules and the substrate. The bonding occurs via a weak interaction known as the van der Waals force (see Advanced Reading Box 3.4). Heating provides enough energy to break bonds between Si atoms at the surface, and the electrons from these form strong covalent bonds with the C_{60}. These are the types B and C molecules in Fig. 3.12, with type B forming covalent bonds with atoms in the top Si layer only and type C forming covalent bonds with atoms in the first two layers. The reason that the images from the three types appear to be different sizes is that types B and C sit lower on the surface. As we will see in Chapter 4, scanning tunneling microscopy is a complex process that does not necessarily give a simple visual image; for example, the apparent width of a feature depends on its height. In addition, the image is more accurately described as maps of electron states, so the bright regions correspond to where the states are most dense. For example, while the Si atoms in the substrate are visible as spots, the carbon atoms are not visible on the C_{60} cages. Instead, one observes features that map the distribution of electrons within the molecule. Note how these features are different in the three types of adsorbed molecule, reflecting the change in electronic structure for the three types of bonding. The features in type A are closest to the free molecule.

This ability to map electronic states is one of the powerful features of scanning tunneling microscopy. By changing the voltage between the probe tip and the sample, it becomes possible to determine how the states are distributed in energy at a given position—a measurement known as scanning tunneling spectroscopy (see Chapter 4, Section 4.4.3). In an atom the electrons occupy discrete energy levels, whose values determine the behavior of the atom. The same is true of a C_{60} molecule that can be considered as an artificial atom containing 360 electrons. Although many of these occupy levels at the same energy ("degenerate states"), there are still a lot of discrete energy levels. There are only two or three, however, that are really important as far as behavior is concerned. In fact there are an infinite number of states if one includes all energies, but in the ground state only the 360 with the lowest energy will be occupied by electrons. All the rest will be unoccupied unless the molecule is given sufficient energy to "excite" an electron from an unoccupied to an occupied state. Chemists dub electron states "orbitals," and the most important orbitals are the highest occupied molecular orbital (HOMO) and the lowest unoccupied molecular orbital (LUMO). In C_{60} the next highest unoccupied orbital to the LUMO (dubbed LUMO + 1) also plays a part. The HOMO–LUMO energy gap determines the minimum amount of energy required to excite an electron in the molecule. For example, the HOMO–LUMO gap in a free C_{60} molecule is 1.75 eV (Table 3.1), which corresponds to a photon wavelength of 700 nm (deep red). Thus a free C_{60} molecule will strongly absorb light of this wavelength.

Table 3.1 lists the HOMO–LUMO gaps for fullerenes adsorbed on various surfaces and with film thicknesses ranging from sparse isolated molecules (such

Table 3.1   HOMO–LUMO Gaps for Fullerenes on Various Surfaces

Molecule	HOMO–LUMO Gap (eV)	HOMO–LUMO + 1 Gap (eV)	Reference
Free C_{60}[a]	1.75	2.78	11
C_{60} on Si(100)[b] isolated molecules	1.8	2.7	12
C_{60} on Ag(001) isolated molecules	1.9-2.1	3.3	13, 14
C_{60} on Au(111) isolated molecules	2.7	3.9	14
C_{60} on Au(887)[c] 0.5–1 monolayer	2.7	3.9	15
C_{60} on polycrystalline Au several layers	2.6	Not measured	16
C_{84} on polycrystalline Au several layers	1.13	Not measured	16
Ce@C_{82} on polycrystalline Au several layers	0.88	Not measured	16
Dy@C_{82} on polycrystalline Au several layers	0.86	Not measured	16

[a] The values shown are calculated.
[b] The values were obtained from the type B molecules in Fig. 3.12—that is, molecules covalently bonded to the Si surface.
[c] Au(887) is a stepped surface produced by cutting at a small angle relative to the (111) direction.

as in Fig. 3.12) to several complete layers. For substrates consisting of single crystals the numbers in brackets give the direction of the surface cut, which dictates the atomic arrangement on the surface. In the case of C_{60} on Si(100) the HOMO–LUMO gap was found to be close to that of the isolated molecule. Note that this result was obtained from the type B molecules shown in Fig. 3.12 since the type A were found to move too easily when trying to do spectroscopy with the scanning tunneling microscope tip. A general feature observed in Table 3.1 is that the HOMO–LUMO gap increases when fullerenes are deposited onto metals and is highest for Au substrates. The relationship between the HOMO–LUMO gap and the type of bonding is not well understood, but in the case of C_{60} on Au and Ag substrates it has been shown that very small differences in the charge distribution at the bonding site are responsible for the 0.6 eV increase in the gap in going from Ag to Au [14], demonstrating that it is very sensitive to the environment. The gap, however, does seem to be independent of the actual Au surface used and is similar for the Au(111), Au(887) (stepped surface produced by cutting at a small angle relative to the (111) direction), and polycrystalline surfaces.

The HOMO–LUMO gap decreases with the size of the fullerene as shown in Table 3.1, which compares the gaps in C_{60} and C_{84} molecules deposited on Au. It is also changed by including a metal atom within the cage as shown for $Ce@C_{82}$ and $Dy@C_{82}$ endohedral fullerenes. The HOMO–LUMO gap in fullerenes is typical of the electronic bandgap in bulk semiconductors used to build electronic devices and is controllable by changing the fullerene size, choosing the substrate, and including endohedral atoms within the cage. The latter can be considered as the equivalent to "doping" used to control the properties of bulk semiconductors. Thus there is the real possibility of building electronic devices using individual fullerenes. This falls into the subject area of "molecular electronics" (see Chapter 5). It is also possible to build interesting and controllable materials by assembling large numbers of fullerenes as described below.

3.7 MATERIALS PRODUCED BY ASSEMBLING FULLERENES (FULLERITES AND FULLERIDES)

In the 1990s, ways were found to manufacture fullerenes in large quantities (see Chapter 4, Section 4.1.9) and it is now possible to purchase purified C_{60} and C_{70} fullerenes in kilogram quantities for about $10 per gram. With such quantities readily available, interest turned to assembling fullerenes into macroscopic materials; and in the condensed form, assemblies of pure fullerene molecules are known as fullerites. To discuss material properties of condensed fullerenes, we will focus on solids made from C_{60} molecules, which are the most studied but many of the properties are shared by fullerites in general.

The bonding between stable fullerene molecules (that is, those in one of the series described in Section 3.5) is by a van der Waals interaction and is relatively weak and isotropic (see Advanced Reading Box 3.4). At room temperature the crystal structure of a C_{60} fullerite, illustrated in Fig. 3.13, is a face-centered cubic (*fcc*) arrangement, which provides the densest possible packing of spheres and is what one would obtain if a pile of ball bearings were uniformly squeezed together. A similar structure is found for other materials whose constituent atoms or molecules interact via van der Waals forces—for example, rare gases. The *fcc* structure is also common amongst metals such as Cu, whose atomic interaction is stronger but is also isotropic. Visually C_{60} soot, whose granules are clumps of C_{60} molecules sticking together, appears as a finely divided black powder, and a piece of the solid large enough to see is black.

Despite the weak interaction, molecules of C_{60} will condense into the solid *fcc* structure at room temperature if, for example, they are vacuum deposited on a surface as shown in Fig. 3.9, but the cages spin freely in their lattice positions. On cooling, fullerite exhibits two structural phase transitions. Below $-13°C$ (260 K) the crystal structure distorts slightly from the *fcc* phase[1] and the molecules no longer rotate continuously, but their motion becomes jerky as they jump from one

---

[1]The new phase has a simple cubic lattice with a four-molecule basis.

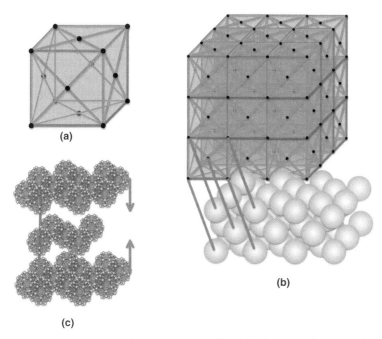

Fig. 3.13 **Face centered cubic structure of C₆₀ fullerite.** (a) The unit cube of the face centered cubic (*fcc*) structure showing the positions in the cube at which to attach atoms or molecules. (b) extended *fcc* structure showing that attaching spheres to lattice sites produces a close-packed structure. (c) *fcc* structure of C₆₀ molecules assembled to produce fullerite expanded in the vertical direction to show how planes come together.

preferred orientation to another. Finally below about $-183°C$ (90 K) the crystal structure remains the same, but the rotary motion of the molecules is frozen into random orientations.

Electrically, fullerite is a semiconductor with a bandgap similar to the HOMO–LUMO gaps found in isolated molecules. Note that some of the results presented in Table 3.1 were obtained from films of fullerenes several layers thick and are therefore measurements for fullerite rather than isolated molecules. Although the fullerite bandgap remains similar to the isolated molecule HOMO–LUMO gap, it is important to understand that the electrons in the corresponding states are not confined to individual molecules. Just as in any bulk semiconductor, the electronic states are extended throughout the entire solid and form the electronic valence and conduction bands. The HOMO state corresponds to the top of the valence band, and the LUMO state corresponds to the bottom of the conduction band. Thus no longer does an individual electron belong to a given molecule, but its wavefunction (probability density) is distributed throughout the entire macroscopic crystal. In a system where the molecules are weakly interacting, these extended states cooperatively produce a charge density around each molecule that does not differ too greatly from that around the isolated molecule, which is why the bandgap remains similar to the HOMO–LUMO gap but there are important differences. For example, in an isolated molecule, an electron excited across the HOMO–LUMO gap—say by heat or absorbing a sufficiently energetic photon—is still bound to the molecule. In fullerite an electron excited across the bandgap becomes a conduction electron and is free to move through the crystal if an electric field is applied. A useful attribute of a fullerite semiconductor is that its bandgap can be controlled from the C_{60} value ($\sim$2 eV) down to zero by choosing the size of the constituent fullerene molecules. It can also be doped by including endohedral fullerenes. Table 3.2 compares the properties of fullerite and silicon. It is evident that fullerite is a light, strong material with a very low coefficient of thermal expansion and a poor thermal conductivity. One important difference between fullerite and conventional semiconductors is that "intrinsic" (i.e., pure, undoped) fullerite is an "n-type" semiconductor [16] (see Advanced Reading Box 3.5) so that the charge carriers are electrons.

ADVANCED READING BOX 3.5—FERMI LEVEL POSITION IN BANDGAP OF FULLERITE

In a solid the Fermi level is the bulk chemical potential of the material—that is, the difference in energy between the N-electron system and the $N + 1$ electron system. In a metal it is the highest energy occupied state in the conduction band, whereas in a semiconductor it is in the bandgap between the valence and conduction bands. It is easy to show that in intrinsic Si the Fermi level lies in the middle of the gap, so the semiconductor is not p-type or n-type. It can be doped with pentavalent atoms to provide an easily ionized electron state close the bottom of the CB, making it n-type, or doped with trivalent atoms to provide an easily ionized hole state just above the VB, making it p-type. In the case of fullerite, measurements [15] have shown that the Fermi level is located above the mid-gap and lies close to the bottom of the CB. The energy separation between the Fermi level and the bottom of the CB is known as the mobility gap and this varies from about 1.1 eV for pure C_{60} fullerite to 0.1 eV for Dy@C_{82} endohedral fullerite. Although new states are not introduced in the bandgap, when the material is connected to a metal, electrons are easily excited from the metal Fermi level into the CB of the semiconductor, making it n-type.

When exposed to intense ultraviolet light, the fullerene molecules in fullerite polymerize; that is, they form strong chemical bonds between adjacent molecules. Pure fullerite will dissolve in toluene but becomes insoluble in its polymerised state. This "photosensitivity" makes it useful as a photoresist in lithographic processes. For example, if a surface is coated in fullerite, one can "write" a pattern onto it with a focused ultraviolet laser or electron beam (see Chapter 4, Section 4.2.1). Immersing the component into toluene will then remove all the fullerite except the regions that have been exposed to the laser.

Table 3.2 Comparison of the Properties of Fullerite and Silicon

Property	Fullerite	Silicon
Crystal structure	*fcc*, lattice constant $= 1.417$ nm	Diamond lattice, lattice constant $= 0.543$ nm
Mass density	1720 kg m^{-3}	2330 kg m^{-3}
Bulk modulus	14 GPa	100 GPa
Volume coefficient of thermal expansion	6.2×10^{-11} K^{-1}	2.6×10^{-6} K^{-1}
Band gap at room temperature	1.7 eV	1.14 eV
Thermal conductivity	0.4 W m^{-1} K^{-1}	150 W m^{-1} K^{-1}
Boiling point	Sublimes at 800 K	3173 K
Resistivity	10^{14} ohm m	10^{-3} ohm m
Electron affinity/atom	2.65 eV	1.39 eV

There are already very good semiconductors and photoresists available, and the advantages gained using fullerites are not sufficient for them to have displaced conventional materials. When other atoms are included in the fullerite lattice to form a compound known as a fulleride, however, the remarkable property of superconductivity can be observed. This is a state achieved below a critical temperature (T_c) in many materials in which the electrical resistance drops to zero; thus, if a current is started in a loop of superconductor, say by induction, then it will flow round the loop forever. The term forever might seem a touch dramatic, but in this case it is valid; at least, it has been estimated that the decay of current in a loop of superconductor would be undetectable by any conceivable instrument over the lifetime of our universe. It has become a holy grail of physics to find a material that is superconducting at room temperature since such a material would revolutionize great swathes of technology and would make levitating trains, for example, easy. So far, that search has proved elusive, and until the 1980s the highest known value of T_c was 23 K in NbGe alloys. In 1986 a new class of superconductors was discovered (the cuprates) with much higher transition temperatures, and the record now stands at 138 K, which is still a long way from room temperature, but there remains some optimism that the holy grail will be found. One of the most exciting aspects of recent discoveries is the range of material that can become superconducting. From 1911 (when superconductivity was discovered) until 1986, all known superconductors were metals but the cuprates are ceramic materials, which are normally insulators, and the more recently discovered fulleride superconductors (in 1991) are normally semiconductors.

As is evident from Table 3.2, fullerite has a high electron affinity; that is, it readily accepts electrons and thus forms compounds with atoms with low electron affinities that tend to donate electrons such as the alkali metals, Na, K, Cs, and Rb. The metal atoms fit into the hollows left between the C_{60} cages after they have been packed into the *fcc* fullerite structure. As shown in Fig. 3.14, for each

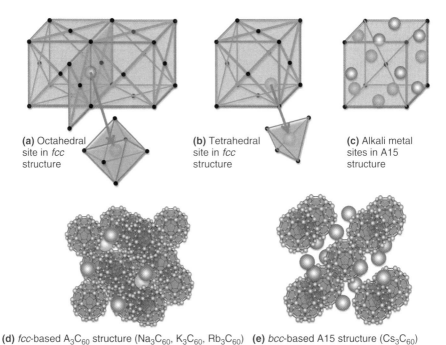

(a) Octahedral site in *fcc* structure   **(b)** Tetrahedral site in *fcc* structure   **(c)** Alkali metal sites in A15 structure

(d) *fcc*-based A_3C_{60} structure (Na_3C_{60}, K_3C_{60}, Rb_3C_{60})   **(e)** *bcc*-based A15 structure (Cs_3C_{60})

Fig. 3.14   Structure of C_{60}—alkali metal fullerides. (a) Octahedral sites for interstitial alkali metal atoms in an *fcc* C_{60} lattice. (b) Tetrahedral sites for interstitial alkali metal atoms in an *fcc* C_{60} lattice. (c) Sites for interstitial alkali metal atoms in a *bcc* C_{60} lattice to produce the A15 structure. (d) A_3C_{60} *fcc*-based structure ("A"=alkali) produced when all octahedral and tetrahedral sites are occupied. This is the structure formed by Na_3C_{60}, K_3C_{60} and Rb_3C_{60}. (e) *bcc*-based A15 structure with the stoichiometry A_3C_{60}. This is the structure formed by Cs_3C_{60}.

C_{60} molecule there is one "octahedral" site and two "tetrahedral" sites that can accommodate extra atoms. Here the names refer to the polygon that is produced by drawing in all the faces produced by triangles of C_{60}. How many of these sites are filled without disrupting the *fcc* structure depends on the size of the filling atom. In the case of Na (diameter 0.429 nm), K (diameter 0.533 nm), and Rb (diameter 0.559 nm), all three sites can be occupied yielding A_3C_{60} compounds (where "A" represents the alkali atom).

As shown in Table 3.3, the superconducting transition temperature increases with increasing size of the alkali atom. Although the interstitial atoms do not disrupt the *fcc* structure, they do push the C_{60} molecules further apart, which in turn increases T_c.[2] Hence the superconducting transition temperatures go in the

---

[2]Fullerides are "heavy fermion" superconductors that rely on a high density of states at the Fermi level. Pushing the C_{60} molecules further apart decreases the width of the valence and conduction bands and increases the density of states at the Fermi level, which stabilizes the superconducting state up to higher temperature.

Table 3.3   Superconducting Transition Temperatures of the Fullerides

Fulleride	Superconducting Transition Temperature (K)	Size of Alkali Atom (nm)
Na_3C_{60}	Not superconducting	0.429
K_3C_{60}	18	0.533
Rb_3C_{60}	29	0.559
$Rb_xCs_{3-x}C_{60}$	33	0.559-0.614
Cs_3C_{60}	38	0.614

sequence 0 K (i.e., not superconducting), 18 K, 29 K in going from Na_3C_{60} to K_3C_{60} to Rb_3C_{60}. K_3C_{60} was the first superconducting fulleride discovered in 1991 [17] with a transition temperature of 18 K; subsequently, attempts to synthesize compounds with the larger alkali atoms Rb and Cs (diameter 0.614 nm) began. Success with Rb_3C_{60} came close behind in the same year [18], yielding a new transition temperature of 29 K. Attempts to produce Cs_3C_{60}, however, were not successful initially. The Cs atom in the full Cs_3C_{60} stoichiometry disrupts the *fcc* lattice and produces a large amount of disorder, although there were hints of small amounts of the material becoming superconducting at a relatively high temperature. In the meantime, inserting a mixture of Rb and Cs atoms into the *fcc* C_{60} lattice did yield a new superconductor with a transition temperature of 33 K [19]. Finally, in 2008 an improved synthesis method produced Cs_3C_{60} in mostly one phase [20], which turned out to be the *bcc*-based "A15" structure shown in Figs. 3.14c and 3.14e. This material currently holds the record superconducting transition temperature for a fulleride (or in fact any organic material) of 38 K.

3.8   DISCOVERY OF CARBON NANOTUBES

As with the fullerenes, the starting point for considering carbon nanotubes is the graphene sheet (Fig. 3.2). Section 3.3 described how introducing pentagonal rings into a graphene sheet will cause it to curl, and 12 pentagonal rings will cause it to close into a pseudo-spherical cage. On the other hand, if we maintain only hexagonal carbon atom rings but roll the sheet into a tube, this is another stable carbon structure known as a carbon nanotube. It is more accurately described as a single-wall carbon nanotube (SWNT), whereas with most manufacturing techniques the tubes mainly consist of several concentric rolled graphene sheets and are multiple-wall carbon nanotubes (MWNTs).

Carbon nanotubes were first identified in electron microscope images by Sumio Iijima in 1991 [21] while working for NEC in Tsukuba. Iijima had previously spotted cage-like fullerenes in images of graphite and sent these to Kroto (one of the discoverers of fullerenes) around 1989 [22]. At the time, these were the only images of fullerenes, and prior to 1989 the structures had been surmised entirely from mass spectra. The initial discovery of tube-like structures was made in an

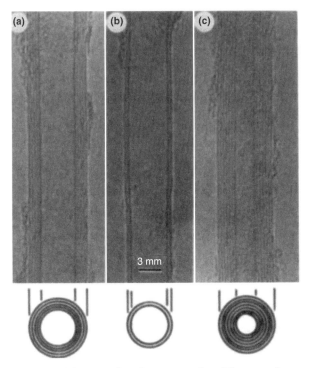

Fig. 3.15 First reported images of carbon nanotubes. Electron microscope images of carbon nanotubes reported for the first time in 1991 by Iijima [21]. These are MWNTs with the number of walls ranging from 2–7. Reproduced with the permission of the Nature Publishing Group from S. Iijima [21].

arc-discharge apparatus similar to the type used to produce beams of fullerenes. Iijima found the nanotubes with diameters ranging from 4 to 30 nm and up to 1 μm in length growing out of the negative carbon electrode. Figure 3.15 shows the first reported electron microscope images of MWNTs, with the number of concentric walls varying from 2 to 7. Iijima reported tubes with up to 50 walls.

The simple arc-discharge method using just carbon electrodes produces large amounts of material, but the tubes that grow are not very controllable in terms of diameter and length. Within two years, Iijima reported an improved technique that included adding an Fe catalyst to one of the electrodes [23]. This promoted the growth of single-wall tubes formed in the gas phase. A range of diameters was found, but two diameters, that is, 0.8 nm and 1.05 nm were dominant, which is reminiscent of the magic numbers found for especially stable fullerene cages. Techniques have steadily improved, and modern methods can produce large numbers of SWNTs with a well-controlled diameter and length with prices dropping toward $50 per gram. It has also been possible to achieve longer lengths, and the world record stands at 18 mm achieved at the University of Cincinatti in

2007 [24]. Although these were relatively large-diameter tubes (20 nm), the length is still 900,000 times the diameter, which, to put it into perspective, corresponds to a 5-mm-diameter household electrical flex with a length of 4.5 km. For the remainder of the chapter the discussion will focus mainly on SWNTs.

3.9 STRUCTURE OF SWNTs

So far the structure has been described simply as a rolled graphene sheet, but there is still a choice of axis about which to roll as shown in Fig. 3.16. Starting from an infinite graphene sheet, one can cut a rectangle and roll the sheet to join at atoms so that the perfect hexagonal arrangement is maintained at a number of different angles. This allows the angle the hexagons make with the tube axis to vary, which may not seem to be a significant change; but as we will see below, the choice of the axis angle produces profound changes in the electronic properties of the tubes. An alternative way of thinking of the rolling angle is to start with a high-symmetry roll, such as cut (a) in Fig. 3.16, and then before

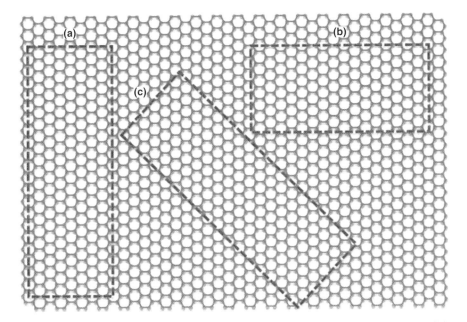

Fig. 3.16 Chirality of nanotubes. Starting with an infinite graphene sheet it is possible to generate families of tubes by cutting rectangles at different angles such as (a), (b) and (c) and rolling up to join at atoms so that the hexagonal lattice is maintained. An alternative way to consider this is to start with a high-symmetry axis such as in (a) and then before rejoining introduce a shift by an integer number of atoms to generate a twist. Each choice of axis determines the chirality of the tube. The chirality determines the electronic properties and for each chirality there is an infinite family of tubes of different diameters and lengths.

joining bonds to atoms introduce a shift of n atoms along the join to generate a twist. The choice of axis angle or n-shift is known as the *chirality* of the tube, and for each chirality there is an infinite family of tubes with different diameters and lengths, which also affect the tube properties. Thus there is a large degree of flexibility in the properties of nanotubes that can be exploited for applications.

The formal method for specifying the chirality of a nanotube is based on the unit vectors of the hexagonal lattice, which are the vectors labeled $\mathbf{a}_1$ and $\mathbf{a}_2$ in Fig. 3.17. It is possible to specify any vector that connects two equivalent points on the graphene lattice in terms of these two by $n\mathbf{a}_1 + m\mathbf{a}_2$, where m and n are integers and several examples are shown. The chirality of a nanotube is specified by the vector $n\mathbf{a}_1 + m\mathbf{a}_2$, with the notation simplified to (m, n). The tube is constructed by rolling the graphene sheet so that the beginning and end of the vectors meet, as indicated by the arrows, Two examples for the nanotubes (8, 8) and (12, 0) are shown in Figs. 3.17b and 3.17c, respectively. It is clear that any (n, n) nanotube will have the "armchair" pattern of carbon atoms at its end highlighted in red in Fig. 3.17b, and any $(n, 0)$ nanotube will have the "zigzag" pattern highlighted in Fig. 3.17c. Sometimes (n, n) and $(n, 0)$ tubes are referred to with these labels. The ends of nanotubes are not necessarily open and will often be terminated by a hemispherical dome, which is half a fullerene.

3.10   ELECTRONIC PROPERTIES OF SWNTs

The discussion of the electronic properties of carbon nanotubes is quite different from that of fullerenes, which are properly regarded as isolated molecules unless they are assembled into a solid fullerite. The length of nanotubes is so much greater than the diameter that they can be regarded as extended wire-like objects, and their individual electronic behavior is better understood in terms of solid-state electronic properties with some peculiarities. As with the atomic structure, the starting point for the discussion of the electronic behavior is the individual graphene sheet, which can be considered to be infinitely extended in two dimensions. Since it has become possible to produce real samples of graphene on which electrical transport measurements can be carried out [25], the electronic properties are no longer just a scientific curiosity and there has been an explosion of interest in this material. It is now understood that it is best characterized as a semi-metal (or alternatively as a zero bandgap semiconductor) with properties intermediate between metals and semiconductors. In a metal the highest occupied electron state (Fermi level) is situated in a continuum of states (conduction band), so there is no discontinuous step in energy required to excite the electron into a higher-energy current-carrying state. Thus applying an electric field smoothly accelerates the electrons, which are then scattered to lower energies but a steady drift current is set up; that is, there is conduction. In a semi-metal like graphene, although there is a continuum of energy states around the Fermi level, the density of electrons goes to zero at the Fermi level. Thus some kind of low-energy excitation such as thermal energy is required to assist conduction. Bulk examples

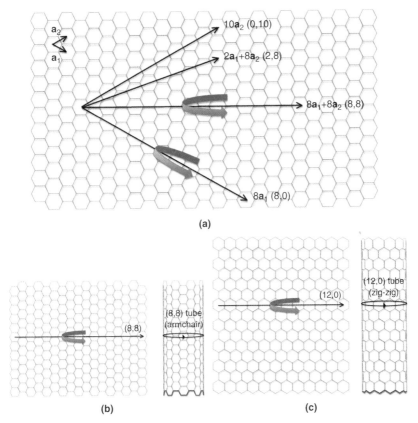

Fig. 3.17 System for specifying nanotube chiralities. (a) The tube is specified by the circumferential vector $n\mathbf{a_1} + m\mathbf{a_2}$ denoting a vector joining two equivalent points on the graphene lattice in terms of the unit vectors $\mathbf{a_1}$ and $\mathbf{a_2}$. The vector notation is simplified to (m,n). The tube is generated by rolling the graphene lattice so that the vector lies on a circumference, as shown by the arrows, and joining the start and end points. (b) An example of an (8,8) tube, also called an armchair tube because of the pattern of carbon atoms at the end. Any (n,n) tube will have an armchair configuration. (c) An example of a (12,0) tube, also called a zigzag tube because of the pattern of carbon atoms at the end. Any $(n,0)$ tube will have a zigzag configuration. For an excellent nanotube modeller that draws the geometry for any chirality see the website: http://jcrystal.com/steffenweber/JAVA/jnano/jnano.html.

of semi-metals are arsenic, antimony, and bismuth. The two-dimensional nature of graphene, however, gives it unusual transport behavior, especially when a magnetic field is applied [26].

When a graphene sheet is rolled to produce a nanotube, which is more like a one-dimensional object, a new quantization condition is set up; that is, the electron waves must have an integral number of wavelengths around the circumference of the tube (Advanced Reading Box 3.6). Thus the electron states only form a continuum in one dimension and are discrete in the orthogonal direction. If one of these states touches the conduction band, the tube is metallic and whether or not this happens depends on the chirality of the tube. It is possible to formulate surprisingly simple rules for the electrical properties of nanotubes; for example, all (n, n) or armchair tubes (Fig. 3.17) are conducting[3] while (n, m) tubes with $n - m = 3i$ (i = integer) are almost metallic but have a very small bandgap generated by the curvature of the tube. Any tube for which $n - m \neq 3i$ is semiconducting with a bandgap that is inversely proportional to the tube diameter. This last result follows simply from the fact that the larger the tube diameter, the closer the spacing of the discrete states and therefore the smaller the gap between the Fermi level and the closest discrete state. Thus carbon nanotubes, which are controllable bandgap semiconductors, have several important potential applications. The use of individual nanotubes as field-effect transistors and other devices is described in Chapter 5, Section 5.4. Also, the carbon nanotube will strongly absorb infrared light whose wavelength matches the bandgap. This attribute can be used to provide localized heating of tissue to treat tumors (see Chapter 6, Section 6.2.4). In this application the controllability of the bandgap becomes important because the absorption must occur at infrared wavelengths at which tissue is transparent.

ADVANCED READING BOX 3.6—ELECTRONIC BANDSTRUCTURE OF GRAPHENE AND NANOTUBES

The electronic bandstructure along k_x and k_y, with energy plotted relative to the Fermi level for two-dimensional graphene, is shown in the figure. Reproduced with the permission of the Nature Publishing Group from P. Avouris et al. [26].

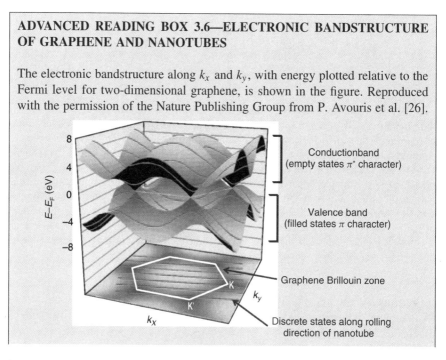

---

[3]Recent work [27] shows that very pure nominally metallic tubes (e.g., (n, n) or armchair tubes) at sufficiently low temperature undergo a type of metal–insulator transformation known as a "Mott transition." At room temperature, however, the tubes are conducting.)

The valence band (filled) touches the conduction band at just 6 points such as those labeled K and K' on the two-dimensional Brillouin zone shown in white. Since there is a continuum of states across the Fermi level, graphene might at first appear to be a metal; but the density of states at these crossing points goes to zero, making it a semi-metal or zero bandgap semiconductor. When the graphene sheet is rolled into a tube along a chiral vector (Fig. 3.17), the two-dimensional continuum of states becomes a one-dimensional continuum in a direction along the tube axis ($k_{//}$). In the orthogonal direction ($k_\perp$) the periodic boundary condition $n\lambda = C$ ($Ck_\perp = 2\pi n$), where C is the tube circumference, λ is the de Broglie wavelength of the electron and n is an integer, gives discrete states indicated by the red lines in the figure. If one of these states crosses the Brillouin zone at one of the K or K' points, there are conduction states available to it and the nanotube is conducting. Otherwise, there is a gap between the filled states and the conduction band, so the nanotube is a semiconductor. The direction of the Brillouin zone relative to the discrete $k_\perp$ states is determined by the chiral vector, so the chirality of the tube determines whether it is metallic or semiconducting. It is possible to formulate surprisingly simple rules for the electrical properties of nanotubes; for example, all (n, n) or armchair tubes (Fig. 3.17) are conducting, while (n, m) tubes with $n - m = 3i$ ($i =$ integer) are almost metallic but have a very small bandgap generated by the curvature of the tube. Any tube for which $n - m \neq 3i$ is semiconducting with a bandgap that is inversely proportional to the tube diameter. This last result follows simply from the fact that the larger the tube diameter, the closer the spacing of the discrete states and therefore the smaller the gap between the Fermi level and the closest discrete state.

3.11 ELECTRONIC TRANSPORT IN CARBON NANOTUBES

A metallic carbon nanotube—for example, an armchair (n, n) tube—can be considered as a very thin wire, but the conduction of charge is very different from a normal macroscopic piece of wire, which, despite its topography, is a three-dimensional object as far as the electrons are concerned. The diameter of a carbon nanotube is of the same order as the wavelength of the electrons, and so the conduction is fundamentally a one-dimensional process and has more in common with light transmitted through an optical fiber than with electrical conduction in a normal wire. The type of conductance is known as ballistic; that is, an electron injected into one end propagates down the tube at about 10^6 ms^{-1} without dissipating any heat. Note that this is different from super-conductivity, which is a cooperative phenomenon in which all the electrons collapse into the same quantum state and when flowing as a current become unstoppable by normal scattering events. In a one-dimensional ballistic conductor, each electron propagates without loss, but resistance is encountered in the contacts at the ends. Also, defects in the conductor can cause additional resistance. Still, the lack of dissipation of heat by the passage of electrons enables one-dimensional conductors to carry huge current densities—even higher than superconductors.

The contact resistance where the nanotube joins the electrical contacts is quantized since the current is only carried in a few quantum states. As discussed in Section 3.10 and Advanced Reading Box 3.6, the conduction band in a nanotube contains only a few states that match onto the continuum of conduction band states in the metal contact. It can be shown that the resistance associated with each electron quantum state (sometimes referred to as a "channel") feeding into a continuum of states is the value $2e^2/h = 12,900$ Ω, where e is the charge on the electron and h is Planck's constant. Two states are like resistors connected in parallel and would halve the contact resistance, and so on. It can also be shown that for any conducting (n, n) or armchair nanotube, there are just two quantum states connecting to the metal contacts, so the contact resistance should always be 6450 Ω for this type of nanotube.

In 1998 de Heer's group at the Georgia Institute of Technology carried out an elegant experiment in which current was passed through various lengths of a single carbon nanotube [28]. This was achieved by contacting one end with the tip of a scanning probe microscope (see Chapter 4, Section 4.4) while the other end was dipped into a low-melting-point liquid metal such as mercury or gallium as depicted in Fig. 3.18. The nanotubes used were MWNT bundles or fibers that emerged from an arc furnace; and as apparent in Fig. 3.18a, there is often a tube that stands proud of the rest, which can be used for the measurement. The tube is lowered using the scanning probe drive into the liquid metal to make contact (Fig. 3.18b) and can be pushed into different depths to alter the length of tube carrying the current. Typically, nanotubes were MWNTs with diameters of the order of 10 nm and lengths of a few micrometers. It was found that the resistance stayed constant at 12,900 Ω, irrespective of the length dipped in the liquid. Clearly, this is the contact resistance arising from a single quantum channel connecting to the contacts. Why there is just a single channel remains a puzzle, but the ballistic nature of conduction in the nanotube is clearly demonstrated. In fact the team went on to show that current densities of up to 10^7 A/cm^2 could be passed through a single tube, which is the equivalent of a household 1-mm Cu mains wire passing 100,000 A! This colossal current density is achieved without any heating of the tube and is 100 times higher than achievable in a superconductor. Above a critical magnetic field, superconductors revert to normal conductors. So although they conduct without any heating, above a critical current density of the order of 10^5 A/cm^2 they generate a magnetic field higher than the critical one and the superconducting state is lost. It should be realized that carbon nanotubes cannot be used to carry very large currents on a macroscopic distance scale because at some point they need to connect to a conventional conductor that must be able to handle the current. In addition, the discussion has focused on short lengths of pristine tube. A long nanotube would have a large number of defects, so there would be dissipation and the electrical transport would not be ballistic.

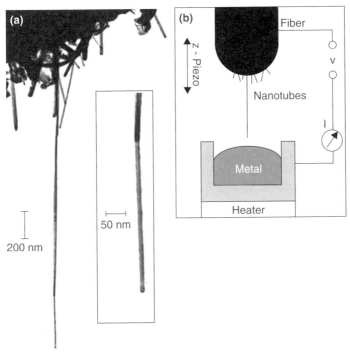

Fig. 3.18 Measuring the resistance of individual nanotubes. (a) TEM image of a fiber consisting of a bundle of MWNT's produced by an arc furnace. It is common to find a single nanotube standing proud of the rest that can be used for measuring electrical transport in a single tube (b) Schematic of experimental set-up. The fiber is attached to the tip of a scanning probe microscope and the SPM controls are used to position the tube with nanometer-scale precision. The tube can be lowered into a container of a low melting point liquid metal (e.g. mercury or gallium) and the resistance can be measured as a function of tube length. The system was used to demonstrate ballistic electron transport in carbon nanotubes, that is, the resistance is independent of length. Reproduced with the permission of the American Association for the Advancement of Science (AAAS) from S. Frank et al. [28].

However, individual carbon nanotubes would be able to carry useful amounts of current ($\sim\mu$A each) around a nanoscale device. Their use in electronics, however, could well emerge before truly nanoscale devices are available. In conventional Si-based integrated circuits the demand for ever smaller components and thinner Cu wires between the components (interconnects) is producing a serious heating problem. The thickness of the Cu interconnects is approaching 100 nm; and it is not just that the thinner wires have a higher resistance, but it becomes more difficult to maintain uniformity in the wire without faults and hot spots. Major chip manufacturers such as Intel are actively researching the use of carbon nanotubes as interconnects.

3.12 MECHANICAL PROPERTIES OF NANOTUBES

In tension, carbon nanotubes are the strongest materials known. The strength arises from the strong covalent sp^2 bonds among the carbon atoms (see Section 3.1) in a graphene sheet, and it is not much different in nanotubes. There are important quantities that define mechanical strength—that is, the tensile (or Young's) modulus and the tensile strength. In a wire the tensile modulus is given by the stress, or applied force per unit area, divided by the strain or the fractional extension of the wire. The strain is a dimensionless fraction, and the stress has the same units as pressure (N/m^2 or pascal—abbreviated Pa). So, for example, if 1000 N (or 100 kg weight) is applied to a steel wire of 1-mm^2 cross section, which causes it to lengthen by 1%, the stress would be $1000/10^{-6} = 10^9$ Pa and the strain would be 0.01, so the tensile modulus would be 10^{11} Pa or 100 GPa. The tensile strength is the force per unit area at breaking point, so again this has units of Pa.

For both these parameters the theoretical values for carbon nanotubes are truly impressive. For example, Young's modulus values are predicted to be about 1000 GPa and tensile strengths could be as high as 300 GPa. These are to be compared with the respective values $\sim$200 and $\sim$1 for steel. Experimental measurements are demanding because to get the values fundamental to nanotubes, stresses have to be applied to individual nanotubes and the strain needs to be measured. The first such experiment was carried out in 2000 on MWNTs produced by an arc furnace [29]. The bundles of tubes were loaded into a scanning electron microscope (SEM) along with an atomic force microscope stage with two facing tips (Fig. 3.19). Single nanotubes could be picked out of the bundle while observing with the electron microscope. The tubes had to be attached firmly to the tips, and this was achieved by using the electron beam to deposit carbon around each end of the tube. It is well known that in an SEM, exposure of a surface to the electron beam for any length of time deposits carbon by cracking contaminant hydrocarbon molecules in the residual vacuum of the instrument just above the surface. In this way the two ends of the tube were "welded" onto the AFM tips. Then under the control of the AFM stage the tube could be stressed by a controlled amount and its extension could be measured.

The measured values for the Young's modulus in this experiment varied from $\sim$270 Pa to about $\sim$950 Pa, so the upper end of the range is close to the theoretical value. The tensile strengths varied in the range 20–63 GPa and thus were significantly lower than the maximum predicted value of 300, but even a single missing carbon atom in the tube can lower the tensile strength by 30%. It is worth noting that even the weakest measured tube was still 20 times stronger than steel.

In compression, carbon nanotubes kink and buckle as one would expect, but they show remarkable resilience in returning to their original state. The obvious mechanical application is in making ultra-strong cables, and an often-quoted application is the "space elevator" that could hoist a load into orbit or launch it

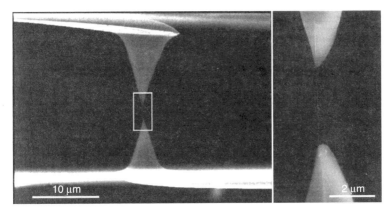

Fig. 3.19   Tensile strength measurements on individual MWNTs. SEM images of individual carbon nanotubes picked up between AFM tips of an AFM stage placed inside a scanning electron microscope (SEM) [29]. The tubes were welded to the tips at each end by using the electron beam to deposit carbon from hydrocarbons in the residual vacuum. Controlled stresses were applied up to the breaking point of the nanotubes and the strains were measured from the images. Reproduced with the permission of the American Association for the Advancement of Science (AAAS) from M. F. Yu et al. [29].

into deep space without requiring rockets. The basic idea is to have a vertical cable thousands of miles long tethered to the ground with a weight at the upper end. The centripetal force acting outwards on the weight and gravity acting on the rest of the cable would keep it under tension, and elevators could climb up the cable depositing payloads into stable orbits at suitable heights. It is easy to calculate that the tensile stress on the cable would be more than 60 GPa—that is, 60 times the tensile strength of steel—so prior to carbon nanotubes there were no materials known that would make the space elevator feasible. The measured values for the strongest individual carbon nanotubes do achieve the required strength; however, there is no known way to maintain these very high values in a macroscopic rope composed of large numbers of carbon nanotubes. Still, knowing that the required tensile strength can be achieved in principle has ignited a good deal of discussion on the web about space elevators as a casual Google search will reveal.

The enormous strength of individual nanotubes could be exploited at the nanoscale by using them to cut softer materials such as biological cells so that their internals can be imaged. This is an important process for biologists who want to obtain high-resolution TEM images of intracellular organelles (see Chapter 6). To obtain these images, the cell must first be sliced into sections thin enough to be electron transparent; each slice is imaged, and the images are stacked to obtain the three-dimensional structure. Normally, these thin slices are obtained by fast-freezing the cell to minimize the damage by ice crystals, and then while maintaining the cryogenic temperature the hardened cell is cut using a diamond

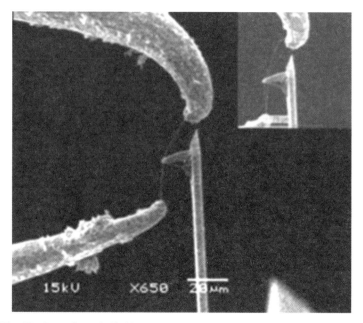

Fig. 3.20   Testing of an individual MWNT for use as a 'nano-cheesewire.' SEM image of Individual MWNT welded to tungsten tips using electron beam induced carbon deposition in an SEM as in Fig. 3.19. The force up to breaking point was measured by pushing an AFM cantilever against the tube. The failure was normally at the welds (inset shows the device at point of failure) but forces approaching 1 μN just before failure were demonstrated. This is sufficient for the device to be used as a 'nanotome' able to slice biological cells into 100 nm slices for TEM imaging (see text). Reproduced with the permission of the Institute of Physics from G. Singh et al. [30].

or glass knife (microtome) into slices less than 100 nm thick. This is a remarkable achievement for a mechanical device; but although it is possible to obtain sufficiently sharp edges for the cut, any knife structure inevitably peels the slices away at an angle and curves them so that the segments tend to crack when laid flat. The solution is to replace the knife morphology by a cheese wire structure so that only the cutting tip is present. This application for individual MWNTs was tested recently by Gurpreet Singh and colleagues at the University of Colorado at Boulder [30] (Fig. 3.20). They placed fine tungsten tips in an SEM and welded the nanotube to the tips using the same technique as the experiment shown in Fig. 3.19 [29]—that is, carbon deposition induced by the electron beam. The strength of individual MWNTs was tested by pushing an AFM cantilever against the tube as shown in Fig. 3.20. It was found that the welds broke before the tube; but for the best case a force approaching 1 μN was achieved before failure, which is of the same order as the force required for the "nanotome" application. With further development these nano-cheesewires could find a range of applications in biology and engineering.

3.13   THERMAL CONDUCTIVITY OF NANOTUBES

Having observed all the other remarkable properties of nanotubes, it should be no surprise that they also display a huge thermal conductivity. The way that thermal conductivity is defined is illustrated in Fig. 3.21. Consider a bar of area A. If a thermal gradient, G, given by $(T_1 - T_2)/L$ (that is, the temperature drop per unit length), causes a power P to flow through the bar, then the thermal conductivity, K, is given by $P/A = KG$. So the power flowing per unit area is the product of the thermal gradient and the thermal conductivity, whose units are W/mK. For example, Cu has a thermal conductivity of about 400 W/mK at room temperature. So if we poked the end of a 10-mm-diameter Cu bar 100 mm long into boiling water, then assuming that the end we were holding was skin temperature, the temperature gradient would be 7 K/cm and the power transmitted along the bar would be about 21 W. We would not hold it for very long under these circumstances.

In a tightly bound network with stiff bonds such as graphene, one would expect the thermal conductivity to be high because lattice vibrations will be efficiently transmitted. Indeed in graphite, which is a macroscopic stack of graphene layers, the thermal conductivity parallel to the layers is 1800 W/mK but only 8 W/mK in the perpendicular direction (along the c axis). A recent measurement of the thermal conductivity of a single graphene layer using a non-contact method [31] showed a thermal conductivity of about 5000 W/mK. For carbon nanotubes a value of 6600 W/mK has been predicted for a single (10, 10) armchair nanotube [32], and a conductivity of over 3000 W/mK has been demonstrated for a single MWNT [33]. If this value was translated to a bulk material with the dimensions in the example above, the power transmitted along the rods would be 150 W.

In the not-too-distant future it is easy to envisage single carbon nanotubes used in circuits of some kind, and it is clear that they are ideal materials with very

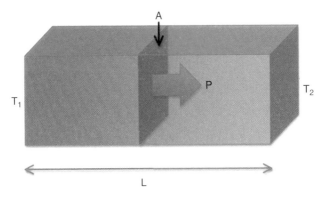

Fig. 3.21   Definition of thermal conductivity. If a temperature gradient $G = (T_1 - T_2)/L$ induces a power P to flow through an area A the thermal conductivity, K, is given by: $P/A = KG$.

high electrical and thermal conductivities. There is also a fundamental interest in their thermal behavior since the quantized lattice vibrations (phonons) are further quantized in the direction around the circumference of the tube because of its small dimensions in a manner analogous to that of the electronic states (Section 3.10). This causes interesting deviations in the low-temperature thermal properties.

3.14 CARBON NANOHORNS

In 1999, Sumio Iijima identified a variation on the nanotube morphology, the so-called nanohorn, in which the tube structure forms a cone terminated by a semi-fullerene cap at one end [34]. These were found in a laser ablation source in the absence of metal catalysts when using a specific power of the laser. It is not surprising to find such a structure since it will arise if a graphene sheet is rolled so that the edges meet at an angle in an analogous fashion to forming a cone from a sheet of paper. At first this may seem a minor variation on the tube structure; but if a carbon nanoparticle source is operating under the right conditions to preferentially make the horns, the larger-scale morphology of the films produced is fundamentally different from that made by tubes. In the case of nanotubes they form either "nano-spaghetti" or parallel bundles of nanotubes like ropes. Nanohorns, on the other hand, pack together at their open ends to form carbon nanoparticles with diameters of tens of nanometers like those shown in Fig. 3.22.

The membrane formed by the graphene sheet is quite holey, and gas can enter the nanoparticles. As a material, this porous stuff presents an enormous surface area per gram to its environment, so it is excellent for soaking up gases such as hydrogen for storage. This is important for future low-carbon-emission energy applications such as fuel cells for road transport. Fuel cell technology is highly advanced, but cheap and safe storage of the hydrogen in the vehicle remains a technological hurdle. Expensive and potentially hazardous solutions such as cryogenic storage of liquid hydrogen have been used in prototype vehicles, but a cheaper method is required for mass produced vehicles. A favorite solution is storage of hydrogen gas at ambient pressure in a highly porous material that releases it on gentle on heating. The critical parameter here is the amount of gas that can be stored per kilogram of the "sponge," and carbon nanohorn particles with their light weight and huge capacity outperform any previously available material.

3.15 CARBON NANOBUDS AND PEA PODS

Before leaving carbon nanostructures, it is worth mentioning that there are some more exotic structures that can be formed including so-called "nanobuds" [35], which are fullerenes covalently bonded to the walls of carbon nanotubes resembling buds on a stalk (Fig.3.23a). These are found when both SWNTs and

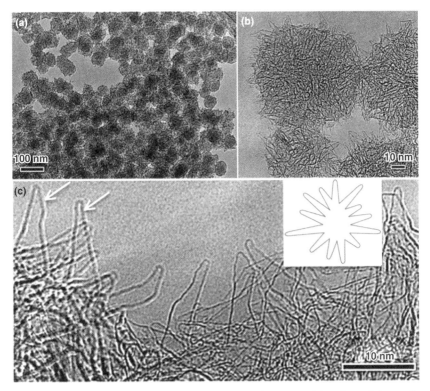

Fig. 3.22   Carbon nanohorns. (a) Morphology of nanoparticles formed by carbon nanohorns. (b) Higher magnification image of nanoparticles. (c) Higher magnification still showing individual nanohorns (arrowed). The ends of the nanohorns are terminated with semi-fullerene caps. The inset shows a simplified schematic of how the horns aggregate to produce the nanoparticles. Reproduced with the permission of Elsevier Science from S. Iijima et al. [34].

fullerenes are produced in the same reaction chamber in the presence of Fe catalyst nanoparticles (see Chapter 4, Section 4.1.10).

Another novel structure is the so-called peapod when a carbon nanotube is filled with fullerenes (Fig. 3.23b). In this case the structure is produced after separate synthesis of the nanotubes and fullerenes [36]. As mentioned in Section 3.9, nanotubes usually have closed ends terminated by a hemispherical fullerene. Prior to filling, these end caps have to be removed, which can be achieved by annealing the tubes at temperatures ~500°C. The "opened" nanotubes and fullerenes are then placed together in a vacuum furnace, which is heated to 200°C–600°C for periods of up to several days. As discussed earlier in the chapter, at elevated temperatures the fullerenes form a vapor; and if one is sufficiently patient, the vapor-phase fullerene will eventually fill the tube. As observed in Fig. 3.23b, dense packing of the tube can be achieved; and with a sufficient vapor pressure of fullerene, it is possible to pack in quite tight-fitting fullerenes.

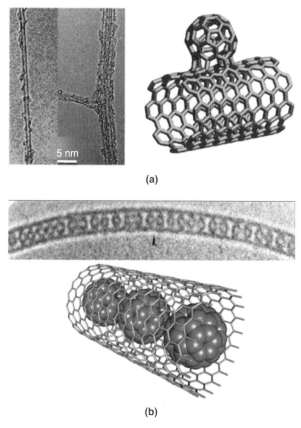

(a)

(b)

Fig. 3.23   Nanobuds and peapods. (a) Nanobud structure (TEM image and schematic) observed when fullerenes and SWNT's are formed in the same reaction chamber in the presence of Fe catalyst nanoparticles. The structure consists of fullerenes covalently bonded to the nanotubes. Reproduced with the permission of the Nature Publishing Group from A. G. Nasibulin et al. [35]. (b) Peapod structure (TEM image and schematic) obtained by filling a SWNT with C_{60} fullerenes post synthesis (see text). TEM image reproduced with permission of Elsevier from B. W. Smith et al. [36]. Schematic reproduced from web page: http://www.chimica.unipd.it/enzo.menna/pubblica/naphod.html with permission from Dr. Enzo Menna, University of Padova.

At present the applications of these hybrid fullerene/nanotube structures are not well-defined, but coupling the flexibility in electronic structure of carbon nanotubes as a function of their chirality with the range of properties associated with fullerenes is bound to produce useful behavior. The fullerenes used to fill nanotubes can also be endohedral fullerenes, themselves encapsulating an atom or molecule and increasing the flexibility of the material still further.

There are other structural variations still—for example, multi-walled versions of fullerenes called "nano-onions" and branched "Y"-shaped nanotubes. Quite possibly, more forms will be discovered. Considering the diversity of structures and properties, it is clear that carbon nanostructures are a very important set of tools within the nanotechnology toolkit.

PROBLEMS

1. Prove that the entire series of spherical fullerenes containing $60 + (k \times 6)$ atoms, where $k = 0, 2, 3, 4, \ldots$, etc., satisfies Euler's Theorem.

2. Given that a carbon atom can be assumed to have a diameter of 0.22 nm, estimate the diameter of a C_{60} molecule.

3. A beam of endohedral fullerenes consisting of C_{60} cages enclosing Na atoms is passed through a laser beam that evaporates C_2 atom pairs from the fullerene cages in a "shrink-wrapping" experiment. Given that the Na atom diameter is 0.429 nm, estimate the number of carbon atoms in the smallest fullerene cage that can enclose the Na atom. Use the carbon atom diameter given in problem 2.

4. Given that assemblies of C_{60} fullerenes condense into a face-centered cubic (*fcc*) crystal structure, use your estimate of the size of the fullerene molecule from problem 2 to estimate the density in kg/m^3 of fullerite.

5. The resistance of a defect-free single-walled carbon nanotube is given entirely by the quantized contact resistance of 12.9 kΩ for a single conducting electron state. Compare this to the resistance of:

(a) A 100-nm-diameter Cu interconnecting wire of length 100 μm used in a high-density integrated circuit.

(b) A Cu wire with a diameter of 2 nm (same as the nanotube) and a length of 100 μm.

The resistivity of Cu is $1.8 \times 10^{-8}\ \Omega \cdot m$.

6. Estimate the percentage reduction in the tensile modulus of a 1.2-nm-diameter single-walled carbon nanotube caused by removing a single carbon atom.

7. A nano-guitar string is made from a defect-free single-walled carbon nanotube (SWNT) of diameter 2 nm and clamped at both ends. One end can be moved with nanometer precision by a piezoelectric device to alter the tension and is adjusted to produce a stress of 100 GPa on the string, at which the vibrating length is 10 μm. The fundamental frequency, f, of a stretched string is given by

$$f = \frac{1}{2L}\sqrt{\frac{T}{m}},$$

where L is the length of the string, T is the tension in N, and m is the mass per unit length of the string. Estimate the fundamental frequency of the stretched SWNT using the diameter of carbon atoms given in problem 2 to determine m.

REFERENCES

1. J. H. Warner, M. H. Rümmeli, L. Ge, T. Gemming, B. Montanari, N. M. Harrison, B. Büchner, and G. A. D. Briggs, Structural transformations in graphene studiedwith high spatial and temporal resolution, *Nature Nanotechnol.* **4** (2009), 500–504.

2. H. Kroto, Symmetry, Space, Stars and C_{60}, *Rev. Mod. Phys.* **69** (1997), 703–722.

3. E. Osawa, Superaromaticity, *Kagaku* **25** (1970), 854–863 (in Japanese).

4. D. E. H. Jones, Ariadne, *New Sci.* **32** (1966), 245.

5. J. Y. Huang, F. Ding, K. Jiao, and B. I. Yakobson, Real time microscopy, kinetics, and mechanism of giant fullerene evaporation, *Phys. Rev. Lett.* **99** (2007), 175503.

6. http://uk.youtube.com/watch?v=NSNlE8AreeM

7. S. C. O'Brien, J. R. Heath, R. F. Curl, and R. E. Smalley, Photophysics of buckminsterfullerene and other carbon cluster ions, *J. Chem. Phys.* **88** (1988), 220.

8. J. R. Heath, S. C. O'Brien, Q. Zhang, Y. Liu, R. F. Curl, H. W. Kroto, F. K. Tittel, and R. E. Smalley, Lanthanum complexes of spheroidal carbon shells, *J. Am. Chem. Soc.* **107** (1985), 7779.

9. A. Kaplan, Y. Manor, A. Bekkerman, B. Tsipinyuk, and E. Kolodney, Implanting atomic ions into surface adsorbed fullerenes: The single collision formation and emission of $Cs@C_{60}^{+}$ and $Cs@C_{70}^{+}$, *Int. J. Mass Spectrom.* **228** (2003), 1055–1065.

10. N. Troullier and J. L. Martins, Structural and electronic properties of C_{60}, *Phys. Rev. B* **46** (1992), 1754–1765.

11. X. Lu, M. Grobis, K. H. Khoo, S. G. Louie, and M. F. Crommie, Spatially mapping the spectral density of a single C_{60} molecule, *Phys. Rev. Lett.* **90** (2003), 096802.

12. X. Yao, T. G. Ruskell, R. K. Workman, D. Sarid, and D. Chen, Scanning tunnelling microscopy and spectroscopy of individual C_{60} molecules on Si(100)-(2×1) surfaces, *Surf. Sci.* **366** (1996), L743–L749.

13. M. Grobis, X. Lu, and M. F. Crommie, Local electronic properties of a molecular monolayer: C_{60} on Ag(001), *Phys. Rev. B* **66** (2002), 161408(R).

14. X. Lu, M. Grobis, K. H. Khoo, S. G. Louie, and M. F. Crommie, Charge transfer and screening in individual C_{60} molecules on metal substrates: A scanning tunnelling spectroscopy and theoretical study, *Phys. Rev. B* **70** (2004), 115418.

15. F. Schiller, M. Ruiz-Osés, J. E. Ortega, P. Segovia, J. Martinez-Blanco, B. P. Doyle, V. Pérez-Diests, J. Lobo, N. Néel, R. Berndt, and J. Kröger, Electronic structure of C_{60} on Au(887), *J. Chem. Phys.* **125** (2006), 144719.

16. M. Shiraishi, K. Shibata, R. Maruyama, and M. Ata, Electronic structure of fullerenes and metallofullerenes studied by surface potential analysis, *Phys. Rev. B* **68** (2003), 235414.

17. A. F. Hebard, M. J. Rosseinsky, R. C. Haddon, D. W. Murphy, S. H. Glarum, T. T. M. Palstra, A. P. Ramirez, and A. R. Kortan, Superconductivity at 18 K in potassium-doped C_{60}, *Nature* **350** (1991), 600–601.

18. K. Holczer, O. Klein, S.-M. Huang, R. B. Kaner, K.-J. Fu, R. L. Whetten, and F. Deidrich, Alkali-Fulleride Superconductors: Synthesis, Composition and Diamagnetic Shielding, *Science* **252** (1991), 1154–1157.

19. K. Tanigaki, T. W. Ebbesen, S. Saito, J. Mizuki, J. S. Tsai, Y. Kubo, and S. Kuroshima, Superconductivity at 33K in $Cs_xRb_yC_{60}$, *Nature* **352** (1991), 222.

20. A. Y. Ganin, Y. Takabayashi, Y. Z. Khimyak, S. Margadonna, A. Tamai, M. J. Rosseinsky, and K. Prassides, Bulk superconductivity at 38 K in a molecular system, *Nature Mater*. **7** (2008), 367–371.

21. S. Iijima, Helical microtubules of graphitic carbon, *Nature* **354** (1991), 56–58.

22. Interview with Sumio Iijima, *Nature Nanotechnol*. **2** (2007), 50–591.

23. S. Iijima and T. Ichihashi, Single-shell carbon nanotubes of 1nm diameter, *Nature* **363** (1993), 603–605.

24. See National Science Foundation (NSF) news website, press release 07–055, http://www.nsf.gov/news/news summ.jsp?cntn id=108992.

25. Y.-W. Tan, Y. Zhang, K. Bolotin, Y. Zhao, S. Adam, E. H. Hwang, S. Das Sarma, H. L. Stormer, and P. Kim, Measurement of scattering rate and minimum conductivity in graphene, *Phys. Rev. Lett.*, **99** (2007), 246803.

26. P. Avouris, Z. Chen, and V. Perebeinos, Carbon-based electronics, *Nature Nanotechnol*. **2** (2007), 605–615.

27. V. V. Deshpande, B. Chandra, R. Caldwell, D. S. Novikov, J. Hone, and M. Bockrath, Mott insulating state in ultraclean carbon nanotubes, *Science* **323** (2009), 106–110.

28. S. Frank, P. Poncharal, Z. L. Wang, and W. A. de Heer, Carbon nanotube quantum resistors, *Science* **280** (1998), 1744–1746.

29. M.-F. Yu, O. Lourie, M. J. Dyer, K. Moloni, T. F. Kelly, and R. S. Ruoff, Strength and breaking mechanism of multiwalled carbon nanotubes under tensile load, *Science* **287** (2000), 637–640.

30. G Singh, P Rice, R. L Mahajan, and J. R McIntosh, Fabrication and characterization of a carbon nanotube-based nanoknife, *Nanotechnology* **20** (2009), 095701.

31. A. A. Balandin, S. Ghosh, W. Bao, I. Calizo, D. Teweldebrhan, F. Miao, and C.-N. Lau, Superior thermal conductivity of single-layer graphene, *Nano Lett*. **8** (2008), 902–907.

32. S. Berber, Y.-K. Kwon, and D. Tománek, Unusually high thermal conductivity of carbon nanotubes, *Phys. Rev. Lett*. **84** (2000), 4613–4616.

33. P. Kim, L. Shi, A. Majumdar, and P. L. McEuen, Thermal transport measurements of individual multiwalled manotubes, *Phys. Rev. Lett*. **87** (2001), 215502.

34. S. Iijima, M. Yudasaka, R. Yamada, S. Bandow, K. Suenaga, F. Kokai, K. Takahashi, Nano-aggregates of single-walled graphitic carbon nano-horns, *Chem. Phys. Lett*. **309** (1999), 165–170.

35. A. G. Nasibulin, P. V. Pikhitsa, H. Jiang, D. P. Brown, A. V. Krasheninnikov, A. S. Anisimov, P. Queipo, A. Moisala, D. Gonzalez, G. Lientschnig, A. Hassanien, S. D. Shandakov, G. Lolli, D. E. Resasco, M. Choi, D. Tomanek, and E. I. Kauppinen, A novel hybrid carbon material, *Nature Nanotechnol*. **2** (2007), 156–161.

36. B. W. Smith, M. Monthioux and D. E. Luzzi, Carbon nanotube encapsulated fullerenes: A unique class of hybrid materials, *Chem. Phys. Lett*. **315** (1999), 31–36.

CHAPTER 4

The Nanotechnology Toolkit

A good deal of nanotechnology is about learning how to make various types of nanostructure—for example, nanoparticles, nanotubes, fullerenes, and so on—and some of the most exciting applications are in combining these different types of structure. The other side of the coin, however, is the ability to image them and probe them to study their properties, and this is what has only become possible in the last few decades. One could argue that nanostructures have been used in technology since the invention of Indian ink, which was probably in China around 2700 B.C. and in its simplest form is a suspension of carbon nanoparticles in water. Also, potters have been using glazes and glass stains based on metal nanoparticles for centuries (for example, see Fig. 0.2). What has really changed in recent years is the ability to study nanostructures at the nanometer level, determine the properties of individual nanoparticles, and, as a result, manipulate them to do what we want. Thus modern nanotechnology developed alongside the probes that enable us to image, measure, and manipulate the nanostructures.

In this chapter the various modern techniques that have been developed to produce nanostructures both by bottom-up and top-down methods will be described in the first part of the chapter. The second part is devoted to some of the probes that are used to study and manipulate the structures. To present every available experiment or probe is beyond the scope of this book, so the discussion is limited to the most commonly used methods and instruments.

4.1 MAKING NANOSTRUCTURES USING BOTTOM-UP METHODS

4.1.1 Making Nanoparticles Using Supersaturated Vapor

A number of nanoparticle sources produce particles by mixing an atomic vapor of the material required with an inert gas at much lower temperature to produce a supersaturated vapor. That is, generate the pressure and temperature conditions at which the vapor will condense into particles. The well-known natural example of this is when you exhale on a cold day and the warm water vapor from your lungs encounters the cold dry air and finds itself in an environment where it would

Introduction to Nanoscience and Nanotechnology, by Chris Binns
Copyright © 2010 John Wiley & Sons, Inc.

normally be condensed into a solid or liquid. The gaseous water condenses into liquid drops that are visible as a cloud. The same process produces raindrops in clouds (see Chapter 2, Section 2.3).

The condensation of a supersaturated vapor into particles can be used for any material except, for metal nanoparticles produced thermally, temperatures of up to 2000°C may be required. There are also methods to make cool clouds of metal vapor described below. The important considerations are the same for all materials whichever method is used to produce the vapor; that is, we need to establish conditions in which the pressure of the atomic cloud of material is greater than its vapor pressure at the temperature maintained in the cloud. For example, Cu has a vapor pressure of about 15 mbar at 1200°C; so if we generated a Cu vapor with a temperature 1200°C and a pressure greater than 15 mbar, it would naturally condense into Cu particles. As was discussed in Chapter 2, Section 2.3, the process of condensation generally requires condensation nuclei (see Advanced Reading Box 2.2)—that is, preexisting particles on which the vapor can condense. It was shown there that for a given set of pressure–temperature conditions, there is a critical size of condensed particle above which the particle will naturally grow and below which it will shrink. Under extreme conditions, this critical size can be smaller than an atom in which case the atomic vapor will naturally condense into particles with no further help (see Advanced Reading Box 4.1). Under less extreme conditions the cloud can still self-nucleate since there is a finite probability that enough atoms will simultaneously collide to form a particle larger than the critical size, which is then stable and continues to grow. This is known as homogeneous nucleation, and the probability of it happening decreases rapidly with decreasing number density in the atomic cloud. Generally, homogeneous nucleation, if it is required, is the slowest part of the growth process and presents a bottleneck to particle formation.

ADVANCED READING BOX 4.1—CONDENSATION OF PARTICLES IN SUPERSATURATED VAPOR

From equation (2.5), Advanced Reading Box 2.2 , the critical diameter, d_c, for a stable particle in the vapour is given by

$$d_c = \frac{4\gamma v}{kT \log_e (P/P_0)}, \qquad (4.1)$$

where v is the volume of the particle, γ is its surface tension, P_0 is the vapor pressure of the material at temperature T, and P is the actual pressure. The ratio P/P_0 is the supersaturation ratio and clearly needs to be greater than 1 or no particles can ever form. Replacing v with $\pi d_c^3/6$ in equation (4.1) gives

$$d_c = \sqrt{\frac{3kT \log_e (P/P_0)}{2\pi \gamma}}. \qquad (4.2)$$

So returning to the above example of Cu vapor, for liquid drops the surface tension, γ, is about 1.2 N/m so at a temperature of 1200°C (1473 K) and pressure of 150 mbar ($P/P_0 = 10$), the critical size is about 0.13 nm, which is smaller than a Cu atom. In this case the vapor will spontaneously condense into particles as every collision of Cu atoms produces a stable condensate from the dimer upwards.

In sources that produce nanoparticles by the condensation of a supersaturated vapor, it is found experimentally that the distribution of particle sizes follows a so-called log-normal dependence shown in Fig. 4.1 (see Advanced Reading Box 4.2). This is a peaked distribution skewed toward larger particles.

ADVANCED READING BOX 4.2—LOG-NORMAL PARTICLE SIZE DISTRIBUTION

It is found experimentally that the distribution of particle diameters of condensed particles in a supersaturated vapor is given by [1]

$$f(d) \propto \exp\left(-\frac{((\ln d) - \mu)^2}{2\sigma^2}\right), \qquad (4.3)$$

where μ and σ are, respectively, the mean and standard deviation of $\ln(d)$. Thus the most probable size (peak of the distribution) is given by $\ln(\mu)$ and the

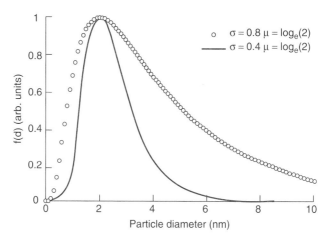

Fig. 4.1 Log-Normal Size Distribution. Distribution of particle diameters formed by condensation in a supersaturated vapor. Both curves have the same most probable size (2 nm) but different width parameters (see advanced reading box 4.2).

standard deviation of particle diameters, σ_d, is given by [2]

$$\sigma_d = \exp\left(\mu + \frac{\sigma^2}{2}\right)(\exp(\sigma^2) - 1)^{1/2}. \qquad (4.4)$$

4.1.2 Sources Producing Nanoparticle Beams in Vacuum

Much of the fundamental research on pure metal nanoparticles is carried out using sources that produce the particles in a supersaturated vapor and transport this through apertures to form beams of particles in vacuum. Due to the high proportion of surface atoms, most materials when they are formed as nanoparticles are very reactive; and if they are to be studied in their pristine state, they have to be produced in an ultra-clean environment. Thus most sources of this type rest at high ($\sim 10^{-7}$ mbar) or ultra-high (10^{-10} mbar) vacuum before the supersaturated vapor is created. To operate the source, a cloud of metal vapor is created by heating or sputtering (see below) and mixed with an inert bath gas such as pure helium or argon in a primary sealed chamber with a single exit aperture (see Fig. 4.2). Conditions for supersaturation are maintained in this chamber at a pressure typically of about 10 mbar, though the pressure in this region can, for some sources, be much higher and above atmospheric pressure. The mixture of condensed particles and bath gas escape from the aperture into a region that is pumped by a very high throughput pump, which maintains a pressure, just outside the aperture, of $\sim 10^{-1}$ mbar. This effusion of gas at high pressure though a small

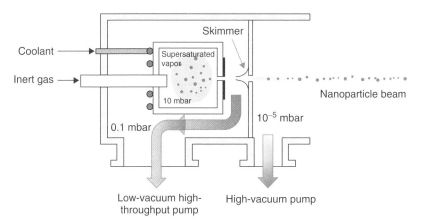

Fig. 4.2 Schematic of Nanoparticle Beam Source. Generic design of a nanoparticle beam source that produces the particles in a supersaturated vapor and forms a beam of free nanoparticles in vacuum. By using suitable apertures and differential pumping the source can maintain a sufficiently high pressure for a supersaturated vapor at one end and have a free particle beam at high vacuum at the other.

hole into vacuum is known as a free jet expansion and can be "strong" or "weak," depending on the ratio of pressures in the high- and low-pressure regions.

The effusing beam encounters a "skimmer" with a small aperture that efficiently transfers the nanoparticles with as little of the bath gas as possible into the next chamber, which is pumped by a high-vacuum pump maintaining a pressure in the high-vacuum region of about 10^{-5} mbar. At this level of vacuum, collisions between the nanoparticles and background gas are rare and the particles are in a beam traveling freely through space.

The particle beam can be mass-filtered (see below), and various experiments can be carried out on the particles in the beam. For example, magnetic nanoparticles can be deflected in a magnetic field gradient to determine their magnetization as a function of particle size (see Fig. 1.7). Alternatively, they can be deposited on a surface and made accessible to a wide range of experimental probes. They can also be imaged using scanning tunneling microscopy (STM) or transmission electron microscopy (TEM) as described in Section 4.4.

A host of sources has been developed around this generic technology, each one specialized to a particular element or particle size range. The main differences between them are the method for producing the metal vapor and the pressures in various parts of the source. Schematics of the different types are shown in Fig. 4.3.

Figure 4.3a shows a seeded supersonic nozzle source (SSNS), which specializes in producing high fluxes of particles of low-melting-point metals such as sodium. The metal is heated by a furnace to a sufficiently high temperature to yield a vapor pressure in the region of 10–100 mbar, and this vapor is mixed with (seeded into) a rare gas introduced at a pressure of several atmospheres. The hot mixture expands into vacuum through a small aperture, and the rapid cooling occurring close to the nozzle condenses the metal into clusters. The clustering continues until the mean free path becomes too long to allow significant interactions between the condensed particles. This is a slight departure from the generic design shown in Fig. 4.2 in that the clustering occurs after the first aperture and the source is characterized by a very strong free jet expansion. The technical difficulty of the containment of the melt in a relatively large furnace has restricted the temperatures achievable to below 1600 K. The source is thus confined to the study of high-vapor-pressure materials. It is, however, capable of producing a flux in excess of 10^{18} atoms/sec of clustered material.

Figure 4.3b shows a common design for a thermal gas aggregation source, which was the first type of metal cluster source reported in 1980 [3]. The metal vapor is produced by a hot crucible, which, with careful design and choice of material (depending on the contained metal), can reach temperatures of over 2000 K; thus a wide range of materials can be produced as nanoparticles, including transition metals. Generally, there is a larger aperture to the high-vacuum region, and the free jet expansion is weaker than in the SSNS. With careful outgassing, the source is particularly suited to producing very clean cluster beams and to UHV-compatible operation with reported vapor pressures of contaminant gases other than the Ar or He bath gas in the 10^{-11}-mbar region [4].

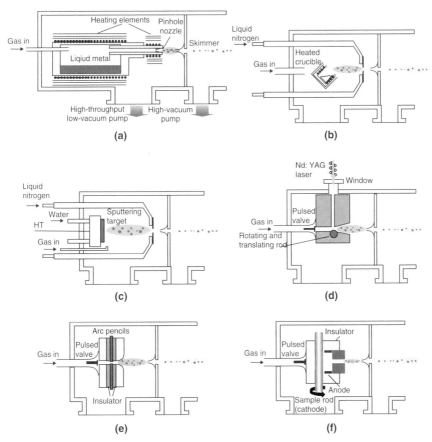

Fig. 4.3   Schematics of the main types of source using super-saturated vapors to produce beams of nanoparticles in vacuum. (a) Seeded Supersonic Nozzle Source. (b) Thermal Gas Aggregation Source. (c) Sputter Gas Aggregation Source. (d) Laser Evaporation Source. (e) Pulsed Arc Cluster Ion Source. (f) Pulsed Micro-Plasma Cluster Source. Reproduced with the permission of Elsevier from C. Binns [15].

An alternative to heating a metal to produce an atomic vapor is sputtering, where high-energy ions are accelerated into a surface and eject atoms from it. The gas that is ionized and used for sputtering is usually Ar with a background pressure that is similar to that of the inert gas introduced into other types of source to produce a supersaturated vapor of the metal. Thus the vapor of sputtered metal atoms condenses into nanoparticles, with the important difference that the vapor is produced in relatively cool conditions. A schematic of the sputter gas aggregation source is shown in Fig. 4.3c. This type of source was originally reported in 1992 [5] and brings several advantages, including the ability to produce clusters of virtually any solid including refactory metals. Clustering is highly efficient because the sputtered vapor consists of not only individual atoms but also a significant proportion of dimers and small clusters, which act as condensation nuclei for nanoparticles. So the start of particle growth does not depend only on homogeneous nucleation overcoming an initial bottleneck for growth. Another characteristic is that a high proportion (up to 50%) of the emerging nanoparticles are ionized, having kept the initial charge donated from sputtering. This is an important factor if the clusters are to be size-selected, since most mass analyzers filter charged particles.

Figure 4.3d shows a schematic of a nanoparticle source that uses a pulsed laser to produce the metal vapor cloud. Light pulses from an Nd-YAG laser focused onto a suitable target can vaporize even refactory materials; and if the laser pulse coincides with a gas burst across the target produced by a pulsed valve, suitable conditions for clustering can be achieved. The first report of this type of source [6] was quickly followed by improvements, including a mechanism for driving the target rod in a screw motion so that a fresh region is exposed to each laser pulse [7]. Clustering occurs within the nozzle as the metal vapor encounters the rare gas and continues in the strong expansion as the mixture is ejected. A useful variable is the phase of the laser pulse relative to the valve opening that can be used to control the size distribution of clusters by altering the average gas pressure during the pulse. Phasing becomes particularly powerful in sources employing two targets to produce binary clusters [8], since it can be used to control the distribution of elements within the cluster. The pulsed output couples efficiently to time of flight mass analyzers (see below).

A pulsed arc can also be used to produce evaporated plumes of metal vapor, and in this case the source is known as a pulsed arc cluster ion source (PACIS) illustrated in Fig. 4.3e [9]. The clustering process is very similar to that found in the laser ablation source. As with the gas aggregation source employing sputtering, the cluster output contains a high proportion of ions ($\sim$10%) and is particularly suited to charged-particle mass analyzers. More recently, a continuous arc source (ACIS) has been developed in which a continuous arc is driven around a hollow cathode by a magnetic field [10]. This is capable of very high fluxes of clusters.

An improved understanding of the operation of cluster sources, achieved by applying fluid dynamics simulations to the gas flows within them, has led to designs that actively control the velocity field of the bath gas to lead to conditions that optimize the cluster flux. An example is the development of the pulsed microplasma cluster source [11] (Fig. 4.3f), which combines design characteristics of several of the other sources. It employs pulses of bath gas directed as jets against the target cathode, which is made of the material to be formed into clusters. The vapor plume is produced by sputtering with the electric discharge also applied as a pulse synchronous with the gas jet. Simulations show that with the correct design of expansion chamber, there is a high gas pressure confined close to the rod in the ablation discharge that increases the sputtering yield. The high pressure also promotes direct sputtering of whole clusters that seed further nanoparticle growth. With a repetition rate of 5 Hz the source has produced carbon nanoparticle films at a rate of 100 μm/hr, making it suitable for coating and device applications.

4.1.3   Mass Selection of Charged Nanoparticle Beams in Vacuum

Generally, the requirement for mass selection in the gas phase falls into two different regimes. Very tight mass selection, able to resolve the number of atoms in the nanoparticles, can be used in gas-phase experiments—for example, magnetic deflection measurements, in which the detector counts the number of incident particles and very low particle fluxes are measurable. On the other hand, in deposition experiments in which measurements are made on some property of the nanoparticle film—for example, the magnetization—large numbers of clusters are required and very high-resolution filters would produce too small an output flux. In this case a lower-resolution filter that passes a greater flux is needed. Ideally, a filter in which the resolution is easily adjustable so that one can choose between flux and resolution is required. In many experiments involving measurements on the macroscopic properties of the films, the samples are produced using the full unfiltered output of the source. As shown in Fig. 4.1, the unfiltered size distribution can be quite narrow, and altering the source conditions such as the bath gas pressure and temperature can control the average size.

Most mass separators require the clusters to be charged; and with the exception of sources that naturally produce a high proportion of ions, such as the sputtering and laser sources, a separate ionizer is required. The simplest design is the electron impact ionizer, which passes high-energy electrons through the cluster beam and ionizes ∼5% of the incident nanoparticles.

If all the charged nanoparticles in a beam are travelling with the same velocity, then mass selection can be achieved by a simple parallel-plate electrostatic deflector as shown in Fig. 4.4a. In general the nanoparticles cannot be assumed to be traveling at the same speed, but in the case of a strong free-jet expansion—that is, a high-pressure differential between the high- and low-pressure regions either side of the first aperture (Fig. 4.2)—the velocity distribution becomes quite narrow. In this situation, if the required mass resolution is not high, simple electrostatic filtering has been used [10].

The deflection angle for particles of mass M is given by the equation in Fig. 4.4a, and it is evident that for a given voltage on the plates the deflection is inversely proportional to M. So a mass-filtered beam can be selected by using a slit, as illustrated in Fig. 4.4a, to exclude all particles but those with a given deflection—that is, a given mass. The operation of the filter is nicely illustrated in Fig. 4.4b, which shows Fe nanoparticles deposited through a simple electrostatic filter onto an Al foil from an ACIS source [10]. This type of source produces almost equal numbers of positively and negatively charged nanoparticles, though the majority remain uncharged. It is clear that films, corresponding to the positively and negatively charged nanoparticles, are deposited either side of the central heavy deposit corresponding to the neutrals. The charged particles have, in addition, been separated according to their mass, with the largest deflections

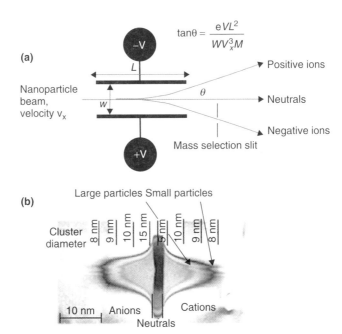

Fig. 4.4 Simple parallel-plate electrostatic filter. (a) Simple parallel-plate electrostatic filter that can be used to separate passes if all the nanoparticles have the same velocity, v_x. The positive and negatively charged particles are deflected either side of the neutral particles by an angle given by the equation and a slit can be used to select particles of a given mass. (b) The nanoparticle beam from an ACIS source passed through a simple electrostatic deflector and deposited onto an Al foil. Either side of the heavy central deposit of neutral particles can be seen deposits due to positively and negatively charged particles deflected by different angles. The particle sizes corresponding to specific deflections are indicated. It is evident that the heaviest of the mass-selected deposits is for particles with diameters of about 8.5 nm. Reproduced with the permission of the European Physical Journal (EPJ) from R. P. Methling et al. [10].

occurring for the lightest particles. The sizes corresponding to specific deflections for the operating parameters used are indicated in the figure. It is clear that the particle size distribution has a peak at a size of about 8.5-nm diameter, indicated by the position in the mass-selected region in which there is the heaviest deposit.

As stated above, the simple electrostatic deflector only deflects a given mass at a specific angle if all the particles are traveling at the same speed. If they have a range of velocities, then a particles with the same mass will be deflected into a range of angles. In practice, nanoparticles will always have a range of velocities, and this limits the resolution of the filter. In many sources the operating conditions produce a wide range of particle velocities, so the electrostatic deflector will not produce a filtering effect at all. If a high mass resolution is required, then the mass selector has to be able to determine the mass independently of the nanoparticle ion velocity. The various types of mass filters developed to do this are described below. The resolving power of a mass filter is the ratio $M/\Delta M$, where M is the central mass passed by the filter and ΔM is the spread of masses around M that are also passed. The resolving power required varies widely, depending on the application. For example, if the nanoparticles are being deposited to produce a material, a filter is only required to produce a narrowing of the native size distribution, so a resolving power of ~ 10 is adequate. On the other hand, for exacting gas-phase experiments that require a knowledge of the number of atoms in the cluster, resolving powers of up to 100,000 are required.

Figure 4.5a shows a schematic of a *quadrupole mass filter*. These have been in use since 1953, mainly at low masses for residual gas analysis in vacuum

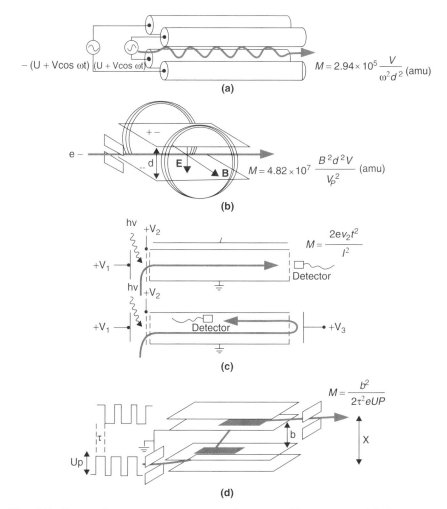

Fig. 4.5   Types of mass spectrometer used to mass filter nanoparticle beams. (a) Quadrupole mass filter. (b) Wien filter. (c) Time of flight mass spectrometer. (d) Pulsed field mass selector [14]. Reproduced with the permission of Elsevier from C. Binns [15].

chambers, but they have some attractive features for filtering nanoparticle beams and have recently been used in gas aggregation sources [12]. The instrument has four parallel cylindrical poles in the geometry shown with opposite poles connected together. A high-frequency voltage with a DC offset ($U + V \cos wt$) is applied to one pair of opposite poles and the negative of the same potential to the other pair. An ion entering the filter undergoes oscillatory motion as it passes through; and for a given set of voltages, U and V, the amplitude is only stable for ions of a given mass, irrespective of their velocity. Ions of the correct mass pass down the axis of the filter oscillating about the center line and exit through the final aperture, whereas for other masses the amplitude of the oscillation grows rapidly and they are flung out of the filter. Although the mass filtered is independent of the ion velocity entering the poles, this does affect the resolution as do the length of the poles, the precision of their alignment, and the frequency of the oscillating voltage. An attractive feature of the instrument is that a simple electronic adjustment (U/V) sets the resolution; so by simply turning a knob, an operator can choose between ion flux and resolving power right up to the resolution limit, given by $U/V = 0.168$. Other advantages are that the filter is light and compact and operates axially, which is convenient in a deposition source. In addition, by lowering the frequency, it is possible to pass arbitrarily large masses. Technically, the maximum resolving power ($M/\Delta M$) obtainable from this type of mass spectrometer is about 4000, though when used for depositing films, it is usually set to much lower resolving powers ~10. The equation in Fig. 4.5a gives the central mass passed at the resolution limit of the filter (when $U/V = 0.168$) in terms of the applied voltages, the ion energy, and the mechanical parameters of the poles.

Figure 4.5b shows a *Wien filter*, which applies electric and magnetic fields oriented perpendicularly to each other and to the path of the charged particles. The fields generate anti-parallel forces on the ions, which balance only for a given velocity irrespective of the mass. So the device is really a velocity filter, and it is operated as a mass filter by initially accelerating the ions through a fixed electrostatic potential, so each mass has a different velocity. The resolution is determined by the collimating slits and the velocity spread in the ion beam prior to acceleration. The influence of the initial velocity spread can be minimized by accelerating the ions up to high speed using a large accelerating potential, but this puts a heavy demand on the magnetic field strength required to operate up to a high mass. In practice the instrument has been demonstrated in a nanoparticle source to filter nanoparticles with up to 1000 atoms with a resolving power of ~20 [13].

The *time-of-flight (TOF) mass spectrometer*, illustrated in Fig. 4.5c, again ensures that different masses have different velocities by accelerating the ions through an electrostatic potential and determines the mass by timing the transition of the ions through the filter. A well-defined pulse of ions is created, by a pulsed ultraviolet laser or pulsed electron impact ionizer. The packet of ions is accelerated into a field-free drift tube, and an ion detector records the arrival time at the end of the tube. The resolution of the simple instrument is limited mainly by the timing accuracy and the pre-ionization spread in position and velocity. These latter two effects can be minimized by the reflectron scheme shown in the figure that uses a simple electrostatic reflector at the end of the drift tube. Slower ions of a given mass take a shorter path in the reflector than do faster ones, so ions of the same mass will be bunched closer together on reflection. This type of mass spectrometer has the highest resolution of those described here, and resolving powers of 100,000 have been reported. The analyzer is not so convenient to use as a filter for cluster deposition, but filtering can be achieved by adding an electrostatic kicker mechanism that pulses a set of steering plates a set time after each ionizing pulse to expel ions of a given mass from the drift tube. This type of mass analyzer is most efficiently matched to a pulsed cluster source such as the PACIS or laser ablation source, each of which naturally produces a pulsed output.

Finally, Fig. 4.5d shows a *Pulsed Field Mass Selector*. This is a simple and ingenious mass filter that uses an electric field pulse to displace an ion beam sideways into a field-free region. As discussed above, if the ions all have the same velocity, they can be separated in mass by a simple pair of electrostatic plates. The pulsed field mass selector exploits the fact that perpendicular to the beam the ions all do have the same velocity—that is, close to zero. For a given strength and timing of the pulse, the lateral velocity given to the ions in the field-free region depends on their mass. A given time later a second decelerating pulse deflects the ions onto a path parallel to the original one and through an aperture. The timing between the lateral accelerating and decelerating pulses determines the selected mass. The instrument can filter masses up to an arbitrarily high limit with a transmission greater than 50%. The resolving power is in the range 20–50, which is adequate for most deposition experiments. The compact inexpensive nature of the filter makes it an attractive option.

Figure 4.6 shows two representative mass spectra at the two extremes of cluster size and mass resolution. A high-resolution mass spectrum was obtained using a time-of-flight mass spectrometer (Fig. 4.5c) of a beam of small Na nanoparticles containing 4–75 atoms (normally called "clusters" when they are this small) emerging from a seeded supersonic nozzle source (Fig. 4.3a) [16]. A feature of this spectrum is that some peaks, indicated by arrows, are particularly intense, especially when compared to the next highest peak. Thus clusters containing 8, 20, 40, and 58 atoms are much more abundant than their neighbors containing an extra atom. These are known as magic numbers and correspond to the number of atoms that produce an especially stable cluster. We came across these in Chapter 3 when discussing fullerenes with a particularly stable morphology. In the region of a nanoparticle source where clusters are growing, each cluster has atoms impinging on it and is also re-evaporating atoms; but if there is an imbalance between the flux on and flux off the cluster in favor of the former, the particle will grow. When it reached the size of a magic number, which is a highly stable configuration, the first additional atom is much more likely to re-evaporate from the cluster surface than one of the atoms in the magic configuration. So if N is a magic number, clusters containing N atoms will be more abundant than those containing $N + 1$ atoms.

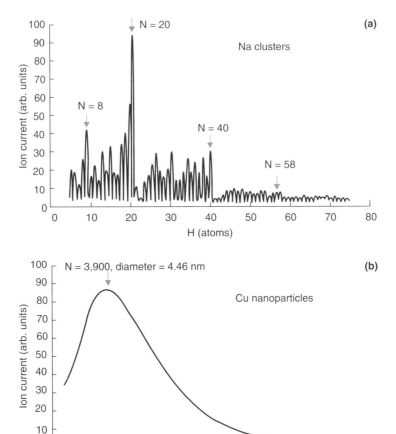

Fig. 4.6 Two extremes of mass spectra. (a) High resolution time of flight mass spectrum [Fig. 4.5(c)] in the size range 4-75 atoms of very small Na nanoparticles (more conventionally called clusters in this size range) emerging from a Seeded Supersonic Nozzle Source [Fig. 4.3(a)] showing a peak for every integer number of atoms and clearly displaying magic numbers (see text). Reprinted with permission from [16]. Copyright 1984 by the American Physical Society. (b) Low resolution spectrum obtained with an ultra-high mass quadrupole mass filter [Fig. 4.5(a) in the range 800-19,000 atoms (2.6 nm–7.6 nm diameter) of Cu nanoparticles emerging from a Sputter Gas Aggregation Source (Fig. 4.3(c))].

In the case of Na clusters, the stability of magic clusters arises from the electronic configuration of the electrons (see Fig. 4.7a). It appears that, to a reasonable approximation, we can forget about the individual Na atoms in the cluster and just consider the whole cluster as a spherical container of electrons, one from each Na atom. Thus a particle made up of N atoms can be considered as a giant atom containing N electrons. Just as in an individual atom, the electron states

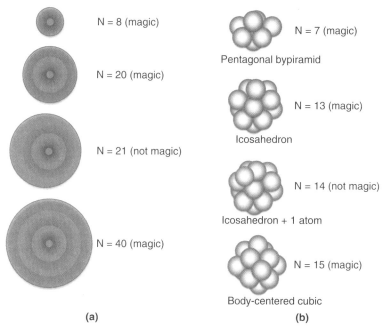

Fig. 4.7 Electronic shell filling and atomic packing origins of magic numbers. (a) Magic numbers produced by electronic shell filling. The cluster can be considered to be a spherical container of electrons, with each atom providing one or more electrons. The electronic states are in a shell structure, just as in an atom and particularly stable structures occur when shells are exactly filled. (b) Magic number sequence produced by efficient atomic packing starting with a pentagonal bipyramid (top) consisting of a pentagon of atoms with additional atoms centred on the pentagon above and below. The next close-packed stable structure ($N = 13$) is the icosahedron consisting of two pentagonal planes oriented perpendicularly to each other enclosing a central atom. Additional atoms are centered above and below the top and bottom pentagons. N=15 is also a magic number (bottom) consisting of a piece cut out from a body-centered cubic structure. If N is incremented by 1 from a magic number as shown for the $N = 21$ and $N = 14$ clusters in (a) and (b) respectively, the additional atom is much more likely to evaporate off so in the environment of a nanoparticle source where the clusters are growing by an imbalance of impinging and evaporating atoms, the magic number N will be found much more frequently than $N + 1$.

of the cluster are grouped into shells and magic numbers correspond to N values required to exactly fill the electronic shells. From a filled shell configuration the energy required to add one more electron (increment the number of atoms in the cluster) and start filling the next highest shell significantly increases the average energy of the electrons in the cluster. The sequence of magic numbers observed in the Na cluster mass spectrum in Fig. 4.6a (i.e., $N = 8, 20, 40, \ldots$) corresponds to the filling of successive electronic shells. A similar sequence of magic numbers is also observed in Cu, Ag, and Au clusters [17].

Electronic shell filling is not the only mechanism for the occurrence of magic numbers; and for some materials, notably the transition metals, magic numbers are due to particularly stable atomic packings around a central atom. For example, time-of-flight mass spectra of small Fe, Ti, Zr, Nb, and Ta clusters all show magic numbers at N = 7, 13, and 15, which correspond to the atomic packings shown in Fig. 4.7b. The basic difference between the electronic shell filling and the atomic packing mechanism for magic numbers is illustrated schematically in Fig. 4.7.

4.1.4   Aerodynamic Lensing and Mass Selection of Neutral Nanoparticles

In the last few years the interaction between researchers working on nanoparticle beam sources and those studying atmospheric aerosols, which were once separate communities, has led to the emergence of new techniques to manipulate nanoparticles in beam sources. An example is the application of aerodynamic focusing illustrated in Fig. 4.8. A series of axial restrictions in the gas flow in a source produces a set of vortices [18, 19] (Fig. 4.8a); and if there are nanoparticles within the gas flow, simulations show that they become increasingly confined about the axis of travel as they travel through the series of apertures [20] (Fig. 4.8b). The effect can be used to boost the flux from cluster sources by (a) inserting the aerodynamic lenses in the high-pressure region of the source just after the clustering is complete and (b) focusing the cluster beam into the skimmer at the start of the high-vacuum region. In addition, the focusing effect is size-dependent and the flow tends to remove the largest and smallest particles from the central beam; thus the focused beam that emerges from the end of the lens system has a narrowed size distribution relative to the newly formed cluster population before the lens (see Advanced Reading Box 4.3).

ADVANCED READING BOX 4.3—MASS-FILTERING USING AERO-DYNAMIC LENSES

The tendency of a small particle to follow an underlying gas flow through an obstacle rather than impact a wall due to its inertia is given by the dimensionless Stokes number (St), defined by

$$St = \frac{\tau v_F}{d},$$

where v_F is the fluid velocity well away from the disturbance produced by the obstacle and d is the characteristic size of the obstacle. The parameter τ is the characteristic time for the gas flow to cause a particle to "swerve" a distance d from its original path. Thus for $St \sim 1$ the particle will tend to clear an obstacle. The figure illustrates the fate of three particles with $St \gg 1$, $St \ll 1$ and $St \sim 1$ in a gas flow through a tube with a constriction. Unless they are travelling

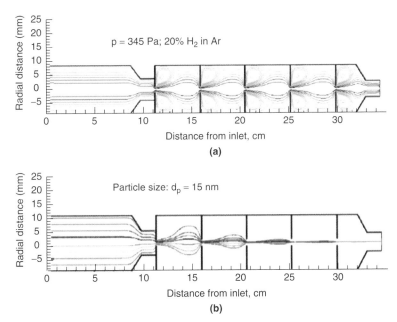

Fig. 4.8 Aerodynamic Lensing. (a) A series of axial restrictions in the gas flow in a nanoparticle beam source introduces a set of vortices. (b) These tend to confine particles in the central beam and after a series of restrictions, particle beam widths of less than 50 μm can be achieved. The same method can be used to achieve partial mass selection since the largest and smallest particles are removed from the beam and a relatively narrow size range passes through (see Advanced Reading Box 4.3). Reproduced with the permission of the Institute of Physics from K. Wegner et al. [20].

along the axis, the largest particles either will crash into the constriction or, if they are deflected through, will not be steered back to forward motion before they hit a wall. The smallest particles will closely follow the gas flow and will get caught up in the vortices after the restriction and deposited on the walls. Only for particle sizes with $St \sim 1$ will the particles be steered through the obstacle and recover their forward motion to travel on to the next one.

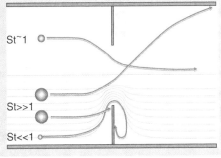

A series of such apertures will progressively remove the largest and smallest particles and filter particles with St $\sim$ 1. The actual particle size that this corresponds to can be adjusted by simply changing the flow conditions (e.g., gas pressure, velocity) and the size of the obstacles

4.1.5   Plasma, Spark, and Flame Metal Aerosol Sources

The nanoparticle sources that produce beams of particles in vacuum described in Section 4.1.2 are best suited to fundamental research. It is possible, using the techniques described in Section 4.1.3, to exercise tight size control-even to the level of choosing the number of atoms required in each particle, and sources capable of working at ultra-high vacuum can generate pristine particles of reactive materials. The main disadvantage is the difficulty of scaling up production to industrial levels. Although there is no reason in principle why it can't be done, the expense of producing, for example, kilogram quantities of nanoparticle material using a vacuum beam source would be prohibitive.

Various methods of producing nanoparticle aerosol in a gas at atmospheric pressure without the need for high-vacuum conditions have been developed, and three are illustrated schematically in Fig. 4.9. As discussed in Chapter 2, if the nanoparticles are contained in a gas at atmospheric pressure, they have a fallout time that can be months (see Advanced Reading Box 2.1), so they are effectively suspended and the gas/nanoparticle mixture can be piped around as if it were a fluid.

A particularly simple and cheap production method is the spark source (Fig. 4.9a) in which a pulsed high-voltage spark is generated between two sharp electrodes made of the material required as nanoparticles. The spark takes place in a flow of inert gas such as He or Ar and is normally produced by charging a capacitor until it reaches a sufficient voltage to initiate a spark and discharge via the plasma generated. Each spark produces a plume of supersaturated metal vapor from the electrodes that condenses into a nanoparticle aerosol in the inert gas. Since it operates at ambient conditions, the source is well-suited to exposing living cells to metal nanoparticles to assess the toxicity as a function of material and particle size. Special care has to be taken with reactive materials because of inevitable impurities in the gas; however, with the use of highly purified gases, clean nanoparticles of Si have been produced with the source [21].

Figure 4.9b illustrates the process used to produce nanoparticles using a thermal plasma source [22]. An inert gas (usually Ar) is mixed with a relatively coarse powder of the material required in nanoparticle form and blown into the coils of a high-power radio-frequency coil that heats the powder-containing gas by induction. The high-frequency field ionizes the gas, which is then induced to carry a heating current. Powers of several kilowatts are used and the powder-containing gas is heated to temperatures of up to 10,000 K and converted into a plasma. Because of the very high temperatures employed, the process can produce

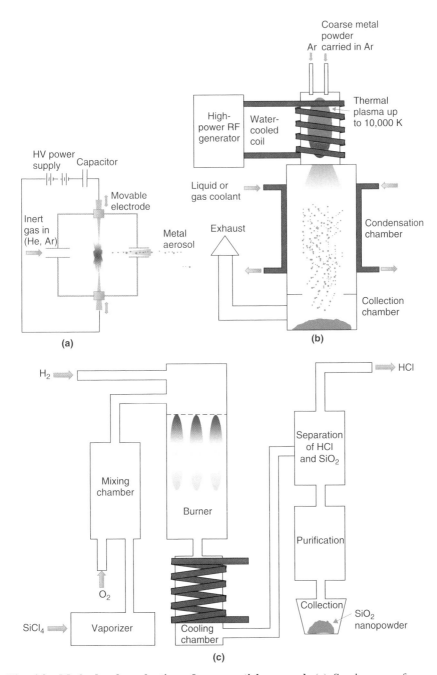

Fig. 4.9 Methods of production of nanoparticle aerosol. (a) Spark source for producing metal or semiconductor aerosol in inert gases. (b) Thermal Plasma source for producing large quantities of metal or ceramic nanopowders. (c) Flame synthesis combined with chemical reactions to produce large quantities of ceramic nanopowders. The method is used for the industrial production of a number of nanopowders and the diagram shows the specific process used to generate SiO_2 particles.

aerosols of just about any material, including refractory metals and ceramics. The plasma is passed into a condensation chamber and is rapidly cooled, either passively by radiation into the water-cooled walls or sometimes by also injecting cold gas into the chamber. The material in the Ar flow condenses into nanoparticles that are collected at the base.

Flames have been used to produce nanoparticle aerosols since prehistoric times, as illustrated in cave paintings; and the most basic demonstration is to smoke a glass plate above a candle flame where the soot deposited is mainly agglomerated carbon nanoparticles. The heat in a flame is generated by a chemical reaction that produces a vapor of reaction products, which stream away from the hot zone and condense into particles as the vapor cools. Flames using appropriate initial combustible materials and mixing them with different gases have produced various types of nanoparticles. As the process has become better understood, good control of the nanoparticle size and composition has been achieved [23]. As an example, Fig. 4.9c shows the industrial process used to make SiO_2 nanoparticles that are used in a diverse range of applications, including fillers in silicone rubber, thickening agents, catalyst carriers, and polishing material. The starting material is liquid $SiCl_4$ that is mixed with oxygen and then hydrogen and passed into a combustion chamber. The flame products, which include HCl and SiO_2, are passed into a cooling chamber where they condense into SiO_2 nanoparticles and hydrochloric acid, which is a commercial by-product of the process. These are separated, and after further purification the SiO_2 nanopowder is collected.

4.1.6 Size Selection of Nanoparticles in Aerosols

If size selection is required in an aerosol, one of the most common filters is a differential mobility analyzer (DMA), illustrated in Fig. 4.10a. The aerosol is introduced into a grounded cylinder, with a high-voltage pole down the middle, along with filtered air to ensure laminar flow. Prior to entry, most of the particles are ionized by a radioactive source such as ^{85}Kr and experience a radial force due to the electric field in the cylinder, which produces a radial velocity that depends on particle size (see Advanced Reading Box 4.4). Only charged particles of the right size develop a trajectory that will pass them through the exit aperture, and the filtered size can be selected by adjusting the pole voltage.

In order to obtain a particle size spectrum, the flux of particles emerging from a filter such as a DMA has to be measured, and detecting small nanoparticles is problematic because they scatter very little light. A common detector employed is a condensation particle counter (CPC) illustrated in Fig. 4.10b. This injects the aerosol and air, again to ensure laminar flow, into a cylinder containing a supersaturated water vapor. The nanoparticles act as condensation nuclei for the water vapor in the tube, and they rapidly increase in size as they travel through. When they emerge, they are large enough to be detected by the scattered light from a laser focused on the particle beam, with each one producing a flash of light at the detector that is recorded as an event. Figure 2.3 (Chapter 2) shows

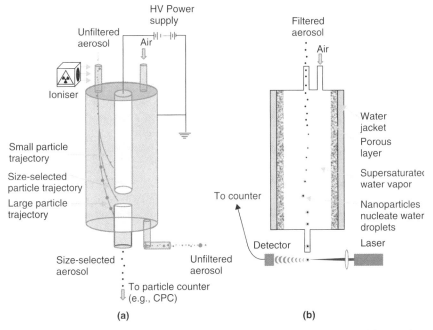

Fig. 4.10   Sizing and counting particles in aerosols. (a) Differential Mobility Analyzer (DMA) that determines particle size in an aerosol. The unfiltered particles are ionized using a radioactive source and then passed into a grounded cylinder with a high voltage pole in the middle. The electrostatic force on the particles produces a constant radial diffusion velocity (see Advanced Reading Box 4.4) that depends on particle size. Only particles of the correct size can escape through the exit aperture. The size can be selected by adjusting the pole voltage. (b) Condensation Particle Counter (CPC) that detects individual particles by passing them through a tube with controlled humidity in which the nanoparticles act as condensation nuclei for water. They can be grown to a sufficient size so that each one produces a detectable flash of light when it passes through a focused laser beam and can be counted. The air introduced into both devices along with the aerosol is to ensure laminar flow.

the size spectrum of carbon nanoparticle aerosol from a candle measured using a DMA and a CPC.

ADVANCED READING BOX 4.4—VELOCITY OF A CHARGED AEROSOL PARTICLE IN AN ELECTRIC FIELD

The drag force on a small particle of diameter d moving through a fluid with viscosity η at a velocity v is given by Stokes' law, that is,

$$F = 6C\pi\eta\,dv,$$

where C is the Cunningham correction for small particles that takes into account that the fluid is not continuous and the spacing of the constituent molecules may be of the same order as the particle size. This correction was mentioned previously in Advanced Reading Box 2.1, which described nanoparticles falling through a fluid medium under gravity. In air, for example, the spacing between the molecules is of the order of 3 nm, so particles of this size or smaller will tend to "slip through" the gaps between the molecules. Also each discrete impact by fluid molecules on the motion of a small particle will produce a significant change in the particle momentum, and the representation by a sphere moving through a continuous medium will become invalid. The significant effect of discrete molecular impacts on small particles leads to Brownian motion.

The important parameter in the Cunningham correction is the ratio of the particle radius, r, to the mean free path of the fluid molecules, λ; that is, $R = r/\lambda$. In terms of R, the Cunninghan correction is given by

$$C = \frac{1}{R}[1.257 + 0.4\exp(-1.1R)].$$

This is plotted in the range $R = 0.1$ to 20 below. In the high R limit, C tends to 1 as required since the motion of large particles approximates closely to motion through a continuous fluid and a correction is not required. For air at room temperature the molecular mean free path is about 150 nm, so only large particles with sizes above ~ 1 μm have C approaching 1. For particles of radius 15 nm the Cunningham correction is ~ 17.

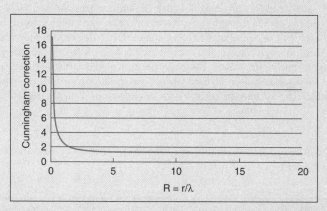

If a particle has a charge of n electrons in an electric field E, its steady-state velocity is

$$v = \frac{EneC}{3\pi\eta d}.$$

This is the radial velocity acquired by a particle in a differential mobility analyzer (DMA).

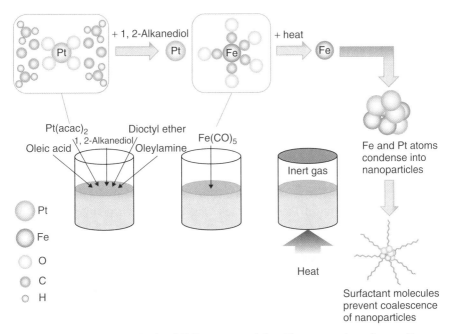

Fig. 4.11   Chemical synthesis of FePt nanoparticles. The preparation of monodisperse FePt nanoparticles using metal-containing pre-cursor molecules Pt(acac)$_2$ and Fe(CO)$_5$.

4.1.7   Chemical Synthesis of Nanoparticles in Liquid Suspensions

For many applications of nanoparticles—for example, medical use (see Chapter 6) or producing ordered arrays of magnetic nanoparticles for data storage (see Chapter 5, Section 5.1)—the nanoparticles are required in a liquid suspension (hydrosol). For this type of material, in general, it is best to use a wet chemical method that naturally produces the nanoparticle assembly as a hydrosol. The generic method is to use metal-containing molecules such as metal salts or organometallics that are chemically reduced to the metal in the presence of surfactants that allow metal clusters to grow but prevent already-formed clusters from agglomerating [24]. To illustrate wet chemical synthesis, the process for producing monodisperse FePt nanoparticles [25], important for future ultra-high density magnetic recording (see Chapter 5, Section 5.1), is described in detail below and shown schematically in Fig. 4.11.

The Pt-containing chemical Pt(acac)$_2$, with the structure shown in Fig. 4.11, is mixed in a reaction flask with the other chemicals. The 1,2-alkanediol is a mild reducing agent that removes the Pt from its molecule, and dioctyl ether is a bulk liquid dispersant with a high boiling point ($\sim$290°C). Small quantities of the long hydrocarbon chain molecules, oleic acid and oleylamine, are also added to form a surfactant around the metal atoms and condensed FePt nanoparticles. To this mixture is added the Fe-containing chemical Fe(CO)$_5$, which is thermally unstable and can be decomposed at relatively low temperature ($\sim$290°C). The mixture is heated to this temperature under an inert atmosphere, and the liberated Fe and Pt atoms condense into nanoparticles that are prevented from coalescing by the surfactant coating. The stoichiometry of the particles can be controlled by the relative amounts of Pt(acac)$_2$ and Fe(CO)$_5$ added, and the size of particles can be controlled in the size range 2–5 nm by the absolute amounts of the metal-bearing chemicals added. Larger particles can also be produced by a two-stage method where small FePt nanoparticles produced in a previous batch are added as seeds into the reaction vessel and the process increases the size of these particles. The method produces highly monodisperse particle suspensions.

An advantage of having the nanoparticles dispersed in a liquid suspension is that producing ordered arrays is relatively straightforward if the particles are monodisperse. A drop of the suspension can be dispensed onto a flat surface; and as the liquid evaporates and the particles come together, the interaction between their surfactant coatings condenses the assembly into an ordered array. In many ways the condensed ordered assembly can be considered as a crystallization of "giant atoms," but there are some important differences. The strong chemical bonds between metal atoms in a normal crystal are replaced by relatively weak bonds between the surfactant coatings around the clusters. The other is that the packing density of the nanoparticles and their arrangement depends on the specific surfactant used. This allows extra degrees of control over the nanoparticle array because the ordered arrangement and the average distance between the particles can be determined by the type and length of the surfactant molecules. It is relatively straightforward to exchange the surfactant coatings in a nanoparticle suspension, so the particles can be formed and the surfactant replaced by another prior to evaporating the liquid. This method of controlling ordered arrays of nanoparticles is illustrated in Fig. 4.12, which shows TEM images of 6-nm-diameter FePt nanoparticles with oleate/oleylamine (Fig. 4.12a) and hexanoate/hexylamine (Fig. 4.12b) coatings. It is clear that the much longer oleate/oleylamine surfactant coatings result in a larger spacing ($\sim$5 nm) between FePt nanoparticles in comparison with the short hexanoate/hexylamine molecules, where the spacing is $\sim$1 nm. In addition, in Fig. 4.12a the nanoparticles are in a hexagonal arrangement, while in Fig. 4.12b they form a square array. This is a powerful degree of control over nanoparticle assemblies.

Another important class of nanomaterials formed by a similar generic method is nanoparticles of semiconducting materials such as CdSe. These are known as quantum dots and fluoresce brightly at wavelengths that depend on the particle size in the range 2–10 nm. There is thus a simple way to control the color of the fluorescence, and simple chemical procedures to synthesize monodisperse dots of a given size have been reported [26]. The Materials Research Science and Engineering Center at the University of Wisconsin has placed a particularly easy-to-follow set of video instructions on the web [27].

An alternative chemical route to producing nanoparticles involves using novel multi-branched molecules known as dendrimers. Deriving their name from the Greek *dendro*, meaning tree, they start from a simple molecule with several

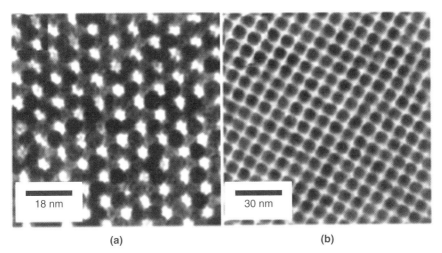

(a)   (b)

Fig. 4.12   Self-ordered arrays of chemically-produced FePt nanoparticles. (a) TEM image of hexagonal array of 6 nm diameter FePt nanoparticles manufactured by the method shown in Fig. 4.11 with an inter-particle spacing of 5 nm produced by coating with oleate/oleylamine. (b) TEM image of square array of 6 nm diameter FePt nanoparticles with an inter-particle spacing of 1 nm produced by coating with hexanoate/hexylamine. Reproduced with the permission of the American Association for the Advancement of Science (AAAS) from S. Sun et al. [25].

active sites onto which similar molecules can be bonded. They can be increased in size by attaching shells of molecules, each with their own active sites that will enable further shells to be added. Each shell is known as a generation, and adding generations increases the size of the dendrimer along with the number of surface active sites. An example of a five-generation polyamidoamine (or G5 PAMAM) dendrimer is shown in Fig. 4.13a. Recently, a method has been developed to use dendrimers as templates to produce metal nanoparticles in suspension with a narrow size distribution [28, 29]. The basic technique is illustrated in Fig. 4.13b; and it starts by adding a metal salt ($AuCl_4$ in the example shown) to a dendrimer solution, causing the dissolved metal ions to permeate the channels of the dendrimer. Next, tetrahydrobiopterin or BH_4 is added and acts as a reducing agent, removing the charge from the metal ions so that they condense into nanoparticles held within the dendrimers. These are labeled dendrimer encapsulated nanoparticles (DENs) and are a uniform size because of the template formed by the dendrimer. To extract the particles, toluene containing a thiol surfactant is added and maintained as a separate phase from the dendrimer solution. Shaking the system causes the thiols to assemble on the nanoparticle surfaces, extract them from the dendrimers, and transport them to the toluene phase. The empty dendrimers are left behind and can be used to manufacture further particles. The particle size (up to a maximum diameter of about 4 nm) can be controlled simply by the amount of metal salt added to the dendrimer solution. The manufacture of

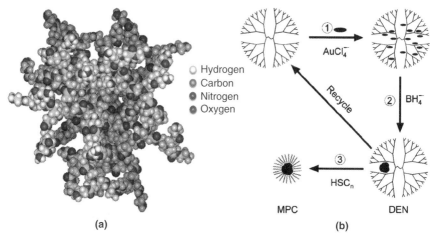

(a)   (b)

Fig. 4.13 Nanoparticle synthesis using dendrimers. (a) G5 PAMAM dendrimer. (Reproduced with permission from Dendritech Inc. http://www.dendritech.com/ pamam.html) (b) Manufacture of Au nanoparticles in dendrimer templates. $AuCl_4$ is added to the dendrimer solution and the dissolved Au ions permeate the interior of the dendrimers. Adding BH_4 reduces the Au and causes it to condenses into a nanoparticle held within the Dendrimer (Dendrimer Encapsulated Nanoparticle or DEN). Adding toluene containing thiols, which remains as a separate phase from the dendrimer solution and shaking causes the thiols to assemble on the surface of the nanoparticle, extract it from the dendrimer and transport it to the toluene phase. The empty dendrimer is then ready to manufacture more nanoparticles. Although there are simpler ways to produce Au nanoparticles the method can be adapted to producing nanoparticles of a number of metals. Reproduced with the permission of the American Chemical Society (ACS) from J. C. Garcia-Martinez and R. M. Crooks [29].

nanoparticles of a number of different metals has been reported with this method [30]. For some applications—for example, catalysis—it is not necessary to extract the metal particle and it can be left in the passivated DEN state. Dendrimers are also used in medical applications as drug carriers (see Chapter 6, Section 6.1.2).

4.1.8 Biological Synthesis of Magnetic Nanoparticles

Certain types of bacteria synthesize nanoparticles of the magnetic oxide magnetite (Fe_3O_4), with each particle enclosed within a biological lipid membrane (see Chapter 6, Section 6.1.2) and the whole structure known as magnetosome. Figure 1.5 shows an electron microscope image of one of these so-called magnetotactic bacteria (*magnetospirillum gryphiswaldense*) with a string of magnetite nanoparticles in its body. The microorganism uses these nanoparticle chains as a simple navigation system to rotate the body of the bacterium along the Earth's field so that it is at a fixed angle relative to the sediment in which is feeds. With evolution having perfected the process over geological timescales, the degree of perfection in the crystal structure and the size uniformity of these nanoparticles is impressive and easily rivals synthetic production methods. Utilizing the bacteria to synthesize magnetic nanoparticles for technological use has a lot going for it, apart from the high quality of the nanoparticles produced. It is environmentally friendly and the magnetosomes are inherently biocompatible, which is important for applications in medicine (see Chapter 6). Nanoparticles synthesized artificially have to be rendered biocompatible by further processing.

The main problem with tapping this source of nanoparticles is that without further intervention the bacteria will only produce magnetite, which is of limited use technologically. In the last few years, effort has been devoted to observing whether feeding the bacteria with minerals containing other metals would induce them to incorporate the metals into the magnetosomes. In 2008 a team from the United Kingdom and France reported that they had successfully induced specific strains of the bacterium *Magnetospirillum* to incorporate up to 1% Co into the magnetite particles [31]. This was an important step forward because even small amounts of Co can produce significant changes in the magnetic properties of the nanoparticles. More recently, a different bacterium was used to synthesize Co ferrite particles with a much higher degree of control over the magnetic properties [32]. This opens the way to commercial biological synthesis of magnetic nanoparticles.

4.1.9 Synthesis of Fullerenes

Small quantities of fullerenes are available from some of the nanoparticle beam sources described in Section 4.1.2 such as the arc, laser evaporation and sputter gas aggregation types. Indeed, as described in Chapter 3, Section 3.2, the discovery and all the early work on fullerenes was carried out using this type of synthesis, which continues in many research groups today. When the interest shifted from fundamental research on individual fullerene molecules to assembling them to produce macroscopic materials, however, methods had to be found to manufacture them cheaply in large quantities. In fact, fullerenes are a natural product and are produced in large quantities by any high-temperature process involving carbon, such as a carbon arc. Designs for carbon arc sources specialized to produce fullerenes have been reported since 1990 [33]. Any large-scale production method inevitably produces a mixture of fullerenes and other carbon structures, so much of the technological development has focused on maximizing the proportion of fullerenes in the soot and separating and purifying specific fullerenes.

Large-scale fullerene production always starts from the same raw material— that is, fullerene-containing carbon black (FB), which is a mixture of fullerenes, amorphous carbon, and hydrocarbons. The preferred method to produce the FB is to use an electric arc between graphite electrodes. The FB or soot from this type of furnace is composed of up to 20% fullerenes by weight. Although it is possible to design the furnace to produce a higher proportion of fullerenes, this is often at the expense of a reduced total FB yield. Within the fullerene fraction, it is mainly C_{60} (65–88%) and C_{70} (12–30%) with the remaining 1–5%

consisting of larger fullerenes. In contrast to other structures composed of carbon, fullerenes are soluble in organic solvents such as toluene, and this is exploited to extract them from the FB. The raw material is mixed with the chosen solvent and then filtered so that the insoluble carbon solids are removed. The enriched fullerene-containing solvent can then be separated into different fullerenes using chromatography [34], a technique that exploits the different rate of passage of dissolved molecules with different sizes through a molecular sieve such as silica gel or alumina. With this method, fullerene purities exceeding 99.5% for C_{60} and 98% for C_{70} can be achieved.

4.1.10 Synthesis of Carbon Nanotubes

As discussed in Chapter 3, the carbon arc furnaces that produce fullerenes also produce large quantities of multi-walled carbon nanotubes found growing from the negative electrode. These have been commercially available for some time, but the challenge has been to develop a method to produce commercial quantities of single-walled carbon nanotubes (SWNTs) with a specified diameter and, if possible, with a controlled chirality. The first step toward large-scale synthesis of SWNTs was taken in 1993 when Sumio Iijima, the discoverer of carbon nanotubes, showed that adding an Fe catalyst to one of the electrodes in the arc discharge apparatus promoted the growth of SWNTs (Ref. 23, Chapter 3).

The three most common generic high-yield methods (Fig. 4.14) to produce carbon nanotubes are the carbon arc method, already discussed (Fig. 4.14a), laser ablation (Fig. 4.14b), and plasma-enhanced chemical vapor deposition (PECVD; Fig. 4.14c). In all cases, employing metal catalysts increases the yield of SWNTs. The laser ablation technique focuses a high-power ($\sim$10 kW/cm^2) Nd YAG laser onto a carbon target, which is vaporized under the laser spot. The hot plume of carbon is generated in a 1200°C oven containing a flow of inert gas such as Ar or He at a pressure of $\sim$0.5 atmospheres. The carbon vapor condenses into a mixture of fullerenes and carbon nanotubes and is carried by the gas flow onto a cooled substrate. Metal loaded into the carbon target also condenses into metal nanoparticles, which form the catalyst for the formation of nanotubes. The particles promote the growth of SWNTs by the mechanism described in Section 4.1.11.

The PECVD method (Fig. 4.14c) generates the carbon vapor by reacting carbon-containing molecules such as acetylene (C_2H_2) with, for example, NH_3. This is done in the presence of a glow discharge plasma (hence the term plasma-enhanced in the name) generated between electrodes, typically by a high-frequency ($>$10 MHz) power supply producing a few hundred volts, although DC can also be used. One of the electrodes is a substrate supporting catalyst nanoparticles on which the nanotubes grow. Often the catalyst is prepared *in situ* by, for example, depositing a continuous metal film and then breaking it up into nanoparticles by ion bombardment. Because of the energy imparted to the carbon atoms in the vapor by the glow discharge, relatively low substrate temperatures down to 300°C can be used. Due to its simplicity,

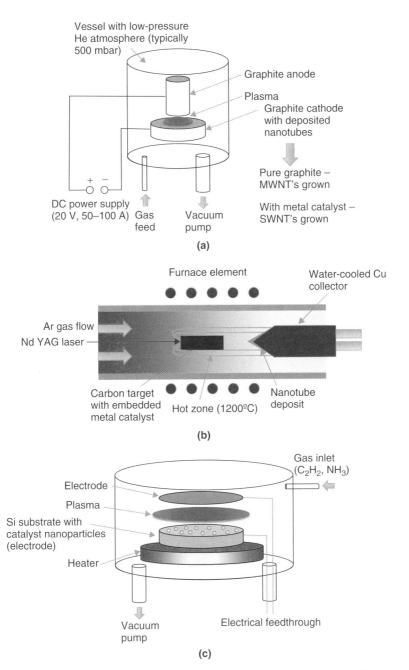

Fig. 4.14 Methods for large scale synthesis of carbon nanotubes. (a) Carbon arc method. (b) Laser ablation method. (c) Plasma-Enhanced Chemical Vapor Deposition (PECVD). In all cases the use of nanostructured transition metal catalysts (usually Fe or Ni) increases the efficiency of production and increases the proportion of SWNT's.

controllability, and relatively low synthesis temperatures, PECVD is rapidly becoming the method of choice for the commercial growth of carbon nanotubes.

One of the useful characteristics of nanotube growth by PECVD is that under the right conditions in the reactor the tubes can all grow vertically out of the surface aligned with each other, much like grass. This is illustrated in Fig. 4.15, which shows some scanning electron microscope (SEM) images of aligned nanotube growth by PECVD taken at the University of York, United Kingdom. In addition, by patterning of the catalyst on the substrate, it is possible to grow the nanotubes just from selected regions as demonstrated in Fig. 4.15c. Structures such as this could make excellent electron emitters for electron guns used in, for example, electron microscopes. A voltage applied to a sharp tip can cause it to directly emit electrons due to the intense electric field developed at the tip, and bundles of nanotubes such as those displayed in Fig. 4.15c form arrays of ultra-sharp tips that can emit high currents at relatively low applied voltages ($\sim$100 V).

The use of patterned catalysts on a surface to promote vertical CNT growth in desired areas is illustrated nicely in the scanning electron microscope (SEM; see Section 4.4.6) image displayed in Fig. 4.16. The Obama faces, each about 0.5 mm across and consisting of about 150 million vertically aligned CNTs, were produced by Professor John Hart at the University of Michigan to mark the victory of Barack Obama in the U.S. presidential election. The "Nanobamas" were synthesized by patterning the catalyst nanoparticles onto a Si surface to follow Obama's face using a process known as photolithography [35] and then growing the CNTs by CVD.

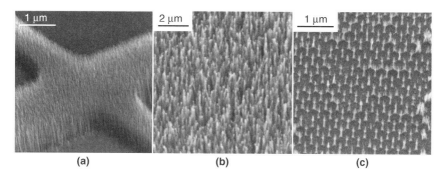

Fig. 4.15   Growth of vertically aligned nanotubes by PECVD. (a) Under the right conditions carbon nanotubes produced by PECVD grow from the catalyst nanoparticles vertically and aligned. (b) Higher magnification view of the vertically aligned structure. (c) By patterning the catalyst the tubes can be grown out of selected regions of the substrate. Images obtained using a Scanning Electron Microscope (SEM- see Section 4.4.6). Reproduced with permission from Prof. M. El Gomati and Dr. Lucy Zhang, Department of Physics, University of York.

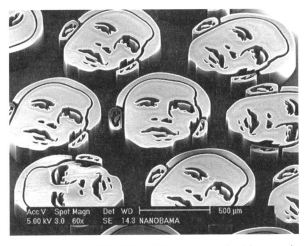

Fig. 4.16   Nanobamas. Face of President Obama synthesized by growing vertical nanotubes on a patterned catalyst on Si. Each face is about 0.5 mm across and contains about 150 million parallel nanotubes (the number of people who voted in the election). Reproduced with permission from Prof. John Hart, University of Michigan.

4.1.11   Controlling the Growth of Single-Wall Carbon Nanotubes

A significant challenge in producing SWNTs for applications in nanotechnology is to achieve control over their diameter and chirality and thus determine their electronic properties. One approach is to select post-synthesis by exploiting differences in the chemistry of nanotubes with different electronic properties. This has been achieved to some extent with simple molecules [36] but has been accomplished more selectively by using sequences of DNA (see Chapter 6, Section 6.1.5) that wrap around tubes with specific chiralities [37]. Once the selected tubes have been tagged by suitable molecules or DNA, they can be separated from the main assembly by, for example, chromatography as used for fullerenes (see Section 4.1.9). The disadvantage of this approach is that it discards the majority of synthesized tubes and it is not conducive to large-scale manufacture, so there is great interest in controlling the synthesis to naturally produce tubes with specific properties. The use of metal catalyst nanoparticles in nanotube manufacture has become standard, and the key to obtaining SWNTs with a controlled structure is a detailed understanding of the mechanism by which carbon nanotubes grow on a metal catalyst particle. In recent years, very high resolution TEM observations with atomic resolution have revealed a detailed picture of how carbon nanotubes grow from metal particles, and some of the clearest images published are shown in Fig. 4.18 [38]. In this experiment the starting material in the TEM was multi-walled carbon nanotubes containing Fe nanoparticles. The sample was heated to $600°C$ *in situ* and irradiated by the electron beam in the microscope with an intensity of 200 A/cm^2. This irradiation dissociated carbon atoms from the MWNT, producing a carbon vapor around the nanoparticles, which seeded

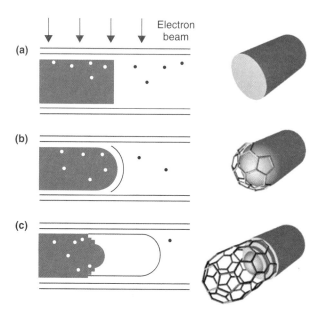

Fig. 4.17 Mechanism of SWNT growth from a metal catalyst nanoparticle. (a) In the experiment described metal nanoparticles within MWNT's are heated to 600°C and irradiated by an electron beam within the TEM creating a carbon vapor around the particles. (b) The end of the particle is covered by a carbon cap and becomes rounded. (c) The nanotube grows from a step formed on the rounded particle surface and carbon atoms diffusing through the body of the particle feed the growth. Reproduced with the permission of the Nature Publishing Group from J. A. Rodríguez-Manzo et al. [38].

the growth of new tubes. Detailed observations of the growth of a SWNT from a particle are shown in the sequence 4.18a–4.18f, and the mechanism is shown schematically in Fig. 4.17 for clarity. In the first stage the carbon atoms enter the body of the Fe particle and also form a cap around the top, producing a rounding of the surface (Fig. 4.17b). The SWNT grows from a step formed in the surface of the curved particle cap, and the growth is fed by carbon atoms diffusing through the body of the particle (Fig. 4.17c).

This mechanism can be seen directly in the images in Fig. 4.18. Figure 4.18a is the onset of growth after the particle has been coated by a carbon cap, and the black arrow indicates the tip of the growing SWNT. It is observed to reach a length of about 8 nm after 15 minutes (Fig. 4.18f). Meanwhile the presence of a metal carbide phase, indicated by the atomic planes (stripes) in the TEM image, is highlighted by the white arrow and is observed to move through the particle. This demonstrates that carbon atoms passing through the catalyst particle feed the growing SWNT.

With a greater understanding of the role of the catalyst, there have been steady improvements in the control over the properties of SWNTs synthesized. They are now commercially available with a diameter variation of only 10%, and the

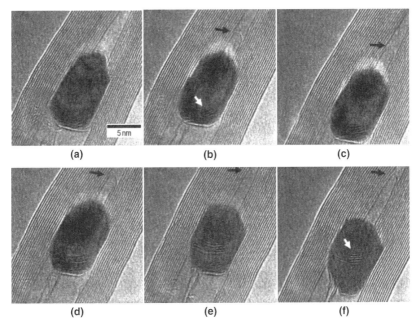

Fig. 4.18   Direct observation of SWNT growing from a metal catalyst nanoparticle.
(a) Onset of growth when the surface of the nanoparticle has been coated with a carbon cap under electron beam irradiation at 600°C. (b)-(f) Sequence of images at 600°C spanning 15 minutes showing the growth of an SWNT whose end is indicated by the black arrow. The white arrow shows the occurrence of a metal carbide phase, indicated by atomic planes (stripes) observed in the TEM image. This carbide phase moves up through the particle showing that the carbon atoms feeding the SWNT growth pass through the catalyst particle. Reproduced with the permission of the Nature Publishing Group from J. A. Rodríguez-Manzo et al. [38].

synthesis can produce yields of up to 50% with the same chirality and up to 90% semiconducting or metallic [39].

4.2   MAKING NANOSTRUCTURES USING TOP-DOWN METHODS

All the nanostructures described so far in this chapter are manufactured "bottom-up"; that is, the conditions are created so that atoms come together in the right way and create the required structures. Since the basic building block of bottom-up methods is a single atom or molecule, they have the ability to produce nanostructures with sizes all the way down to a single atom (0.1–0.3 nm). The main drawback is the lack of flexibility of the type of nanostructures that can be produced, these being restricted to whatever forms naturally within a given environment. On the other hand, the toolkit of bottom-up methods is constantly expanding and becoming more versatile.

Top-down manufacture of nanostructures proceeds by "machining" the structure out of a larger block, in the same way as engineers' tools like lathes and milling machines shape metal blocks to produce mechanical components. The advent of modern tools, such as electron beam lithography, described below, has enabled machining to be carried out on a scale of tens of nanometers. Although the size resolution is not as high as with bottom-up methods, there is huge flexibility in the type of nanostructure that can be produced, and any shape can be manufactured out of just about any material. Bottom-up and top-down methods are thus complementary, with the former offering very high spatial resolution, down to a single atom but limited manufacturing flexibility, and the latter being very versatile in production capability but having a limited spatial resolution. As top-down methods are developed to produce ever-smaller structures and the flexibility of bottom-up methods improves by the discovery of new ways to make nanostructures, there is a convergence between the two. Some of the most exciting developments in nanotechnology are a result of combining top-down and bottom-up methods, so that, for example, a tiny nanoparticle produced by a bottom-up approach can be built into a larger, but still nanoscale, device built by a top-down method (see Section 4.3). In the following sections the most common top-down methods for building nanostructures are described.

4.2.1   Electron Beam Lithography (EBL)

Electron beam lithography (EBL) relies on materials or "resists" that are chemically altered when they are exposed to an electron beam. The chemical change renders them soluble in a solvent in which they were previously insoluble (positive resist) or vice versa; that is, they become insoluble in a solvent in which they are soluble prior to exposure (negative resist). Using electron optics employing electrostatic and magnetic lenses, a high-energy electron beam can be focused to a spot ~5 nm across, which can be rastered across a surface, thereby "writing" soluble or insoluble regions into a resist with a resolution of a few nanometers. The first electron beam resist, the organic material polymethyl methacrylate (PMMA), was discovered in 1968 [40] and is a positive resist when used with the solvent methyl isobutyl ketone (MIBK). This remains the standard resist in use today, though the key to getting yet higher resolutions (that is, the creation of smaller features) is the development of new resists [41].

Figure 4.19 shows how EBL can be used to create metal nanostructures on a surface using positive and negative resists. Starting with the bare substrate, usually Si, a thin layer of resist is coated on the surface (Fig. 4.19b). If PMMA is used, it is dissolved in MIBK and the strength of the solution determines the thickness of the PMMA film. Next (Fig. 4.19c) a highly focused electron beam usually at an energy of about 100 keV and a beam current of a few picoamperes is rastered over the required regions to "write" the desired nanostructure on the surface. The dose must be carefully calculated so that the chemical alteration in the resist is complete and throughout its whole depth. Some resists, including

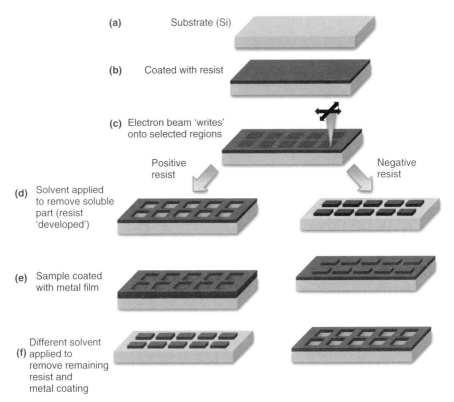

(a) Substrate (Si)

(b) Coated with resist

(c) Electron beam 'writes' onto selected regions

Positive resist

Negative resist

(d) Solvent applied to remove soluble part (resist 'developed')

(e) Sample coated with metal film

(f) Different solvent applied to remove remaining resist and metal coating

Fig. 4.19   Creating metal nanostructures on a Si surface using Electron Beam Lithography (EBL). The bare substrate (a), usually Si, is coated with a layer of resist (b), which can be positive or negative. The desired pattern is 'written' onto the resist by a highly focused electron beam (c) after which the unexposed resist is removed by a suitable solvent (d). The required metal is deposited onto the sample to the correct thickness (e) after which all the remaining resist and its metal coat is washed off with a suitable solvent (f). The process leaves an array of metal islands or an array of holes in the film with the use of positive or negative resists respectively.

PMMA, can be switched from positive to negative by altering the electron exposure; thus at low doses it is a positive resist and becomes negative if given a sufficiently high dose. The resist is then exposed to the solvent, often by simple dipping, and is "developed"; that is, the unwanted resist is dissolved away (Fig. 4.19d). In the example shown, this leaves an array of square holes with bare substrate at the bottom in the case of a positive resist and an array of square islands on a bare substrate with the negative resist. Next the metal required in the nanostructure is deposited on the surface to the correct thickness (Fig. 4.19e). Finally, a solvent that can dissolve unexposed resist is applied; and this lifts off all the metal that is not in contact with the bare substrate (Fig. 4.19f), leaving square metal islands or square holes in a continuous film for the positive and negative resists, respectively.

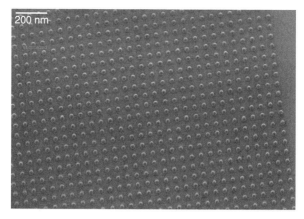

Fig. 4.20   40 nm CoPt magnetic dots produced by EBL. SEM image of array of CoPt dots produced by the process shown in Fig. 4.19 using a PMMA (positive resist) on Si. The CoPt film was deposited at stage Fig. 4.19(e) by sputtering. The array was used as a test of ultra-high density magnetic memory storage in which each magnetic 'dot' stores a single data bit, that is a '1' or a '0' encoded by the magnetization direction (see Chapter 5, Section 5.1). Reproduced with the permission of the American Institute of Physics from V. L. Mironov et al. [42].

Figure 4.20 shows an array of 40-nm-diameter CoPt nanoparticles separated by 80 nm manufactured on a Si surface using EBL. This was used as a test of ultra-high-density magnetic memory storage in which each magnetic "dot" stores a single data bit—that is, a "1" or a "0" encoded by the magnetization direction (see Chapter 5, Section 5.1).

Modern electron optics is capable of focusing the high-energy electron spot to an area of less than 5 nm on the surface, but it is still impossible to create features with this size resolution using EBL due to a number of other limiting factors. The main one is electron scattering within the resist, which spreads the beam within the resist, making the exposed region wider than the focused spot (see Advanced Reading Box 4.5). This can be minimized by using very thin resist layers. Another limiting factor is an observed swelling of the resist layer during the development phase (Fig. 4.19d), which is reduced if the development is carried out in conjunction with ultrasonic stimulation [41].

The state of the art in the spatial resolution of EBL for the production of general patterns is about 20 nm, though individual lines have been produced on surfaces with thicknesses to below 10 nm [41]. This may improve further with the development of new classes of resist.

At present, EBL has the highest spatial resolution of the most common top-down methods and can manufacture nanostructures approaching the size of bottom-up methods, but it has limitations in producing patterned structures over large areas. Typically, the maximum electron beam deflection is 100 μm, so to write patterns over larger areas requires moving the sample stage and tiling 100-μm^2 areas. This puts enormous demands on the precision of the sample movement, which needs to be able to position the stage with nanometer accuracy.

A more serious problem is the time required to produce a patterned nanostructure over a macroscopic area. Within a single 100-μm $\times$ 100-μm area, which can be accessed by the electron beam without moving the sample, the pattern can be written in a few minutes, but to cover a 1-cm^2 area would require a couple of weeks. To write a structure on a 700-cm^2 area on an industrial-sized Si wafer used in the electronic industry would take years. Until this problem is solved, EBL will remain a nanotechnology tool restricted to research laboratories.

ADVANCED READING BOX 4.5—ELECTRON SCATTERING WITHIN RESIST AND BEAM BROADENING

When high-energy electrons enter a material, they undergo inelastic collisions with the electrons within the atoms of the material. These collisions transfer energy to the bound electrons, which are liberated from their parent atoms and become fast electrons traveling through the material, themselves undergoing scattering and creating more secondary electrons. The scattered electrons spread in a cascade starting from the initial point of incidence as illustrated in the figure. The lower the energy of the secondary electrons, the more likely they are to scatter and the more new secondaries they create, but each generation of secondaries has a lower energy until finally they no longer have enough energy to liberate bound electrons and they assimilate themselves back into the bound charge of the material. It is the very process of ionizing the electrons from the bound atoms that produces the chemical change that renders it soluble or insoluble, and so the relevant chemical change is occurring throughout the secondary electron cloud in the material, which is wider than the incident electron beam.

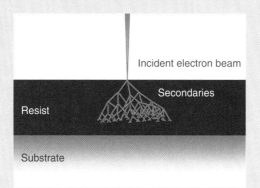

One approach to minimize the broadening due to electron scattering is to use very thin resist layers as illustrated on the next page.

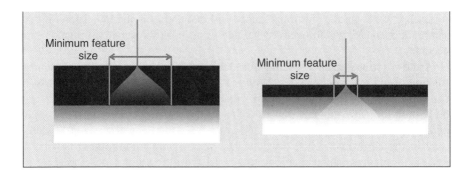

4.2.2 Manufacturing Nanostructures Using Focused Ion Beams (FIB)

As with electrons, beams of ions can also be focused down to spots on a surface <5 nm across. This has been possible using liquid metal ion sources (LMIS) originally developed in the 1970s [43]. The most commonly used liquid metal in modern focused ion beam (FIB) instruments is Ga due to its low melting point (29.8°C), and a schematic of a liquid gallium source is shown in Fig. 4.21. Metal ions are generated from a tiny source just a few nanometers across by an ingenious method, which is also used in electrospray technology. If a conducting liquid surface is subjected to an intense electric field, this distorts the natural shape of the liquid surface formed as a result of surface tension (Fig. 4.21a) and above a critical field value the shape of the liquid forms a "Taylor cone"[44] with a sharp tip. The electric field required to form the Taylor cone ($\sim 10^8$ V/cm) is sufficient to pull atoms off the sharp tip and ionize them, which is the basis of electrospray sources. A video of a Taylor cone forming as the electric field is increased is available on YouTube [45].

The method is implemented in an LMIS by having a sharp tungsten needle emerging from the bath of liquid Ga (Fig. 4.21b). The liquid metal wets the tungsten tip and on applying a sufficient extraction voltage a Taylor cone is formed on the tip of the tungsten needle. The point from which Ga^+ ions are emitted can be <5 nm across, producing an extremely small source, which can be refocused to small spot on the sample surface. As the Ga^+ ions are pulled off the end of the tungsten needle, the liquid flows to replenish it. Figure 4.21c shows an LMIS from a commercial FIB manufactured by the FEI Company. With moderate use, a source can last up to 2 years between replacements.

The focused ion beam at the sample is produced by magnetic lenses in a column shown schematically in Fig. 4.22. On extraction from the LMIS, the ions have an energy in the range 5–50 keV and are focused into a parallel beam by a condenser lens. The diameter of the parallel beam is reduced by a defining aperture, of which there are usually several of various sizes on a carousel. The parallel beam passes down the column and is brought to a sharp focus ($\sim$5 nm at low beam currents and the smallest apertures) at the sample surface by the focusing lens. In addition, there are lenses (usually octupoles) to deflect the beam, along with a beam-blanking electrode to "switch off" the ion beam at the sample.

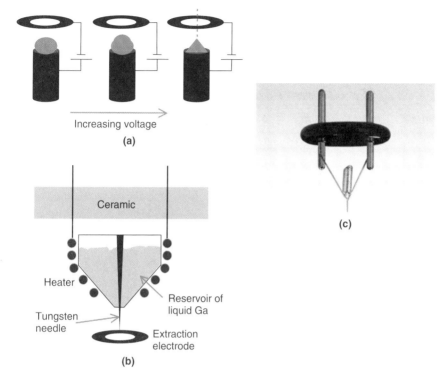

Fig. 4.21   Liquid Metal Ion Source. (a) Formation of Taylor cone on applying a sufficiently large electric field to the surface of a conducting liquid. (b) Schematic of a Liquid Metal Ion Source (LMIS) used in a Focused Ion Beam (FIB) instrument. The liquid metal (usually Ga with a melting point of $29.8°C$) wets the sharp tungsten needle and a high extraction voltage forms a Taylor cone on the tip of the needle whose apex can be less than 5 nm wide. Ga+ ions are field emitted from the tip and replenished by liquid metal from the reservoir. Due to the very small source size the ion beam can be focused to a spot <5 nm in diameter on the sample surface. (c) Photo of an LMIS used in a FIB manufactured by FEI. Reproduced with permission from FEI.

When the ion beam hits the surface, a number of processes occur as shown schematically in Fig. 4.23. It is different from a focused electron beam, described in Section 4.2.1, which can break bonds but for solid surfaces, like metal or Si, is nondestructive. In this case, the mass of the energetic ions is similar to that of the atoms in the surface, which can be directly ejected—a process known as sputtering. In the process the ejected atoms can be ionized, so a mixture of neutral atoms and ions is ejected. Some of the sputtered atoms or ions can land back on the surface, causing unwanted structures around processed regions. In addition to the sputtered atoms, secondary electrons are also generated, and these can be used for imaging. Some of the incident ions burrow into the surface and are implanted a few atomic layers into the material.

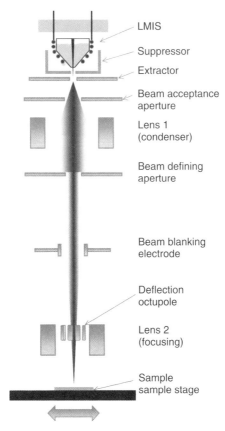

Fig. 4.22   Schematic of ion beam column. The LMIS and extractor provide the small ion source, which is focused into a parallel beam by the condenser (Lens 1). A selectable beam-defining aperture reduces the diameter of the beam and after passing down the column is brought to a sharp focus (<5 nm at low beam currents and small apertures) at the sample by the focusing lens (Lens 2). In addition to the main lenses are beam deflecting lenses (usually octupoles) and a beam-blanking electrode to turn off the beam at the sample.

At very low currents, when sputtering is minimal, the focused ion beam can provide high-resolution (~5 nm) images of the surface by rastering the beam across the surface and measuring the secondary electron yield. Topographical features show up by differences in secondary electron yield. This can also be done with similar resolution and nondestructively by an electron beam as used in an EBL system and is the principle of operation of a scanning electron microscope (SEM; see Section 4.4.6). This is a common laboratory instrument, and some of the images shown in this and other chapters are obtained using an SEM. Despite the potential for surface damage, there are some advantages in imaging using ion beams, including the greater sensitivity of the electron yield to material. Some

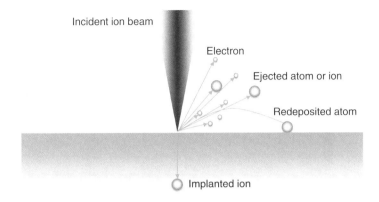

Fig. 4.23 Products of incident ion beam. When the energetic ions hit the surface they eject atoms or ions out of the surface (sputtering) some of which are re-deposited near the incidence point. In addition secondary electrons are generated. Some of the incident ions are implanted a few atomic layers deep in the material.

commercial FIB instruments have a separate electron column installed next to the ion column, so SEM images of a surface can be obtained nondestructively after processing with the ion beam.

One of the main uses for a FIB instrument is "milling," which is the removal of material from a surface by sputtering. Naively, one can think of the beam as a 5-nm scalpel cutting into the surface, but in practice the minimum size of the milled features is tens of nanometers. The loss of resolution is due to the fact that, at currents required for milling, as opposed to imaging, it is no longer possible to focus the beam down to 5 nm. In addition, sputtered atoms themselves sputter more atoms on their way out of the surface, so material is removed from an area larger than the focused beam. Some of the sputtered atoms are redeposited on the surface, resulting in further degradation of the intended structure. On a commercial instrument, there is a user-friendly interface that enables the operator to define areas on a surface to be milled away. The FIB software defines the selected areas in terms of overlapping pixels, whose size depends on the beam conditions and then controls the hardware to land the beam on the defined pixels sequentially. The instrument is highly versatile; and any solid surface, from soft organics to refractory metals, can be milled.

Figure 4.24 shows a nanoscale magnetic AND gate produced using a FIB. The starting substrate was Si coated with a 20-nm-thick Co film and then a 15-nm-thick carbon layer to protect the Co from oxidation. The FIB was used to mill away a pattern in the Co/C film through to the bare Si substrate, leaving the three Co/C islands shown in the figure. If the two "input" islands on the left are both magnetized in one direction, they magnetize the larger island in the same direction. It is evident that a resolution of better than 50 nm has been achieved in the milling of the carbon-capped Co film.

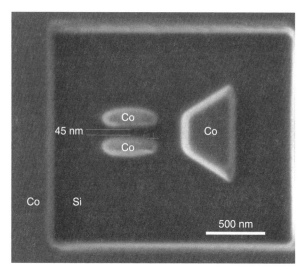

Fig. 4.24   Magnetic AND gate produced by fib milling. Magnetic AND gate patterned into a carbon-capped 20 nm thick Co film on Si. The film has been milled through to the bare Si to produce a three-island structure of Co. If the two "input" islands on the left are both magnetized in one direction, they magnetize the larger island in the same direction. The resolution achieved in the milling of the Co/C film is better than 50 nm. Unpublished image produced by the author using an FEI Quanta 3D dual beam FIB/SEM system.

In general, the resolution of FIB is not as good as EBL and depends on the material being milled. The sputtering rates are very sensitive to the specific element or alloy being machined and are also sensitive to the crystal orientation. Thus in a polycrystalline sample the sputtering rate changes at each crystallite boundary, resulting in an uneven surface. The best resolution $\sim$20–30 nm is achieved on single-crystal Si samples and approaches that of EBL.

Some of the limitations of FIB, including uneven sputtering rates and re-deposition of material, can be reduced by chemically enhancing the sputtering process. This is done by flowing a gas jet over the milled area from a fine tube about the size of a hypodermic needle $\sim$100 μm away as illustrated in Fig. 4.25a. The gas species is chosen to be a substance that adsorbs on the surface; and when the high-energy focused ion beam hits it, the molecules crack, producing a species that reacts with the surface to produces a new material with a higher sputtering yield, which can be an order of magnitude higher than the original surface. The volatile by-products of the reaction are ejected from the sample. For example, the gas Cl_2 flowing over a Si or an Al surface enhances the sputtering yield by a factor of 7–10. It is also possible to depress the sputtering yield with a suitable choice of gas, and H_2O vapor does this for many surfaces.

With a suitable choice of gas, it is also possible to deposit material at the point where the beam is focused (Fig. 4.25b). This time the gas is chosen to be a material that again sticks to the surface, but whose molecules crack into

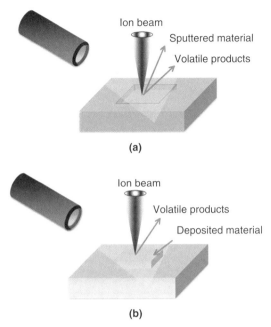

Fig. 4.25 Ion sputtering with precursor gases. (a) Chemically enhanced FIB in which the precursor gas molecules are fragmented by the ion beam at the focal point to form a species that reacts chemically with the surface to produce a material with a higher sputtering yield and volatile species that are ejected. (b) Deposition in which the precursor gas molecules are fragmented by the ion beam at the focal point to produce an inert material that deposits on the surface and volatile species that are ejected.

(i) a species that is unreactive and sits on the surface and (ii) a volatile species that is ejected. The material deposited also needs to have a relatively low sputtering yield so that it is deposited faster than it is sputtered away. A common example is the precursor gas $W(CO)_5$, which decomposes to W that is deposited on the surface and volatile C- and O-containing compounds that are ejected. With suitable precursor gases, it is also possible to deposit Pt, Al, SiO_2, and C [46]. The material deemed to have been deposited is actually the majority material because the deposit is not very pure. For example, when using $W(CO)_5$, the deposit contains 75% W, 10% C, 10% Ga, and 5% O. This is still useful, as a conducting contact for example. The main benefit of FIB-induced deposition is that material is only deposited where the ion beam is incident, so it is possible to "draw" very fine structures on the surface. The spatial resolution for deposition is slightly degraded relative to milling, but it is still possible to build features with a resolution of a few tens of nanometers.

Figure 4.26 shows a comparison of milling and deposition on the same surface. By building up a structure layer by layer, it is also possible to create three-dimensional structures; for example, see Fig. 0.4 in the introduction, which shows "the world's smallest wine glass," produced by FIB deposition of C.

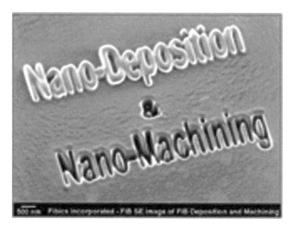

Fig. 4.26 Comparison of milling and deposition using a FIB. SEM image of the same structure milled into and deposited onto a surface using FIB. Reproduced with permission from Fibics.com, Ontario, Canada.

Although in general the resolution of FIB is less than that of EBL, its huge versatility in being able to mill any material and deposit a variety of materials with sub-100-nm resolution has made it an attractive instrument for top-down nanotechnology research. It is especially useful when combining bottom-up and top-down synthesis (see Section 4.3).

In summary, the state of the art in top-down nanotechnology is the ability to make structures with sizes ~10–20 nm, which overlaps at the large end with bottom-up techniques that generally assemble structures smaller than 5 nm. A method in the toolkit that uses Bottom-up and Top-down methods simultaneously to synthesize nanostructures is known as dip pen nanolithography, but the description of this is deferred until after the section on atomic force microscopy (see Section 4.4.5).

4.3 COMBINING BOTTOM-UP AND TOP-DOWN NANOSTRUCTURES

The smallest nanostructures are produced using bottom-up methods and involve the natural self-assembly of atoms or molecules. Examples include carbon nanotubes, fullerenes, and metal nanoparticles (Section 4.1). As discussed above, top-down methods can now produce structures on a length scale not much larger, so it has become feasible to include a tiny nanostructure produced by a bottom-up method within a designed nanoscale device manufactured using a top-down method. The important point is that since top-down methods can produce structures all the way up to the macroscopic scale, we now have the ability to address a single bottom-up particle directly with macroscopic probes such as conductivity measurements.

An illustration of this is Fig. 4.27, which shows an individual multi-walled carbon nanotube draped across four submicron gold electrical contacts produced

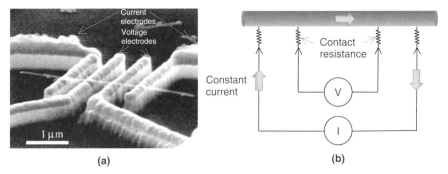

(a)   (b)

Fig. 4.27   4-point probe conductivity measurement on single carbon nanotube. Illustration of combining top-down and bottom-up methods to attach electrodes to an individual multi-walled carbon nanotube. The gold contacts in the SEM image are produced by Electron Beam Lithography. Reproduced with the permission of the Nature Publishing Group from A. Bachtold et al. [48]. (b) Schematic of 4-point probe resistance measurement to eliminate the effect of contact resistance.

by electron beam lithography. A general problem with measuring the electrical transport properties of single nanotubes is the large contact resistance at each electrode that masks the native conductivity of the tube. The traditional way of getting around this with any resistivity measurement including those at the macroscopic scale is to use a four-point probe method illustrated in Fig. 4.27b. A constant current, I, is passed through the outer two contacts, and the voltage dropped across the inner two contacts, V, is measured. Since no significant current flows in the voltage measuring circuit, the resistance, R, of the portion of the sample between the inner two contacts is given by Ohms law, $R = V/I$. Even so, very large contact resistances compared with the resistance being measured can cause problems using a four-point probe. In the specific experiment that used the device shown in Fig. 4.27a, it was found that the contact resistance was reduced significantly when the contact areas were exposed to the electron beam of the SEM [47]. This is due to the "carbon welding" process that was described in Section 3.12, Chapter 3. The four-point resistance measurement of a number of tubes showed resistances in the range $0.35-2.6$ kΩ. It should be emphasized that the conduction in these tubes was not the quantized ballistic conduction observed with defect-free single-walled nanotubes discussed in Section 3.11, Chapter 3. Multi-walled carbon nanotubes with defects produce resistance by electron scattering as in a normal conductor, so the tube in question would behave like a very thin conventional wire. Assuming the length along which the resistance is measured is 0.5 μm and the tube is 20 nm in diameter, Cu and graphite wires of the same diameter would have resistances of $\sim 100\Omega$ and ~ 2 kΩ, respectively.

The device structure shown in Fig. 4.27 was also used to demonstrate a quantum-mechanical effect known as Aharonov–Bohm oscillations (see Advanced Reading Box 4.6). These are periodic variations in the resistance of a tube when a magnetic field is applied along the axis of the tube and is due to the wave properties of the conducting electrons. Electron waves traveling around the circumference of the cylinder interfere and an axial magnetic field changes the phase of the waves and alters the effect of the interference on the conductance.

ADVANCED READING BOX 4.6—AHARONOV–BOHM OSCILLA-TIONS IN A CARBON NANOTUBE

For a conductor that has the morphology of a thin-walled cylinder, as in the figure below, conduction electrons can take a circumferential trajectory like the one labeled p that circles the cylinder once. For each of these, there is an equivalent time-reversed trajectory, p^*, that is in the opposite direction and interferes with p. With no applied magnetic field, the interference is always constructive, which leads to electron backscattering and a positive correction to the resistance (i.e., an increase). This effect is known as weak localization.

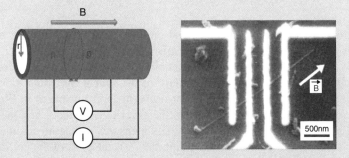

A magnetic field applied along the tube axis changes the phase of the electron waves in the two trajectories and thus their interference. As the field is increased, the interference starts to become destructive, weak localization is lifted, and the resistance drops. When the field is increased beyond a specific value, the interference becomes constructive again, weak localization is restored, and the resistance increases through a maximum. As the field is swept, it can be shown that, for a cylinder with radius r, there is a peak in the resistance whenever the field is

$$B = n\frac{\hbar}{er^2},$$

where $n = 0, 1, 2, \ldots$ is an integer, $\hbar$ is Planck's constant divided by 2π, and e is the electron charge. It is easy to calculate that for a macroscopic cylinder an enormous field would be required to see a single oscillation, but for nanotubes the $n = 1$ oscillation occurs for easily realizable fields as demonstrated by the data below, which were obtained from the single nanotube across electrical contacts shown in the right-hand figure above [48]. It is observed that the

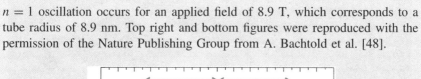

$n = 1$ oscillation occurs for an applied field of 8.9 T, which corresponds to a tube radius of 8.9 nm. Top right and bottom figures were reproduced with the permission of the Nature Publishing Group from A. Bachtold et al. [48].

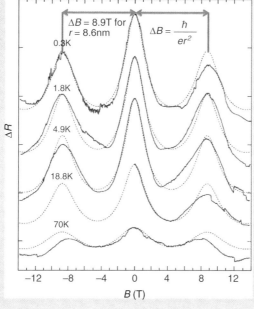

A more sophisticated hybrid device constructed around an individual single-walled carbon nanotube with a diameter of 1 nm is shown in Fig. 4.28a and represented schematically in Fig. 4.28b. This was built by depositing the SWNTs in a water/surfactant suspension onto Si, imaging their location and then patterning the electrodes on top of the nanotube using EBL [49]. In this case the contacts are Al, which becomes a superconductor below 1.2 K, and the configuration of the device is a superconducting quantum interference device (SQUID). This is a loop of superconductor with two breaks in the loop known as Josephson junctions. In the device shown, the Josephson junctions are formed by the strands of SWNT.

In a normal SQUID, the Josephson junctions are thin non-superconducting layers that allow a supercurrent to pass by quantum mechanical tunneling of the superconducting electrons through the barrier that is formed. In such a device the critical current (i.e., the maximum current that can be passed without the superconductor reverting to a normal conductor) is modulated by a magnetic field applied to the device, and it forms a very sensitive magnetometer. The device based on SWNT Josephson junctions shown in Fig. 4.28 has some additional features when compared to a normal SQUID. As discussed in Section 3.11, Chapter 3, a defect-free SWNT only has a small number of discrete states that

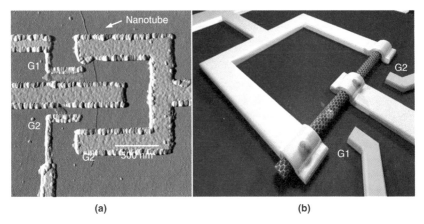

(a)   (b)

Fig. 4.28   SQUID with Josephson junctions formed by a SWNT. (a) Atomic Force Microscope (AFM – see Section 4.4.4) image of superconducting electrodes connected to an individual SWNT using EBL. The grey Al electrodes form a superconducting loop (below 1.2 K) with the SWNT providing two Josephson junctions in the loop to make a SQUID. The gate electrodes marked G1 and G2 with false gold color can be used to independently control each Josephson junction to turn it 'on' or 'off'. Reproduced with the permission of the Nature Publishing Group from J.-P. Cleuziou et al. [49]. (b) Schematic of the device. The Josephson junctions are so sensitive to local magnetic fields that they would be able to detect the magnetization of a single molecule (shown attached to the nanotube). Reproduced with permission from Prof. Wolfgang Wernsdorfer, Laboratiore Louis Néel, Grenoble, France.

can conduct electrons. The voltage on the gate electrodes G1 and G2 can be adjusted to line up or misalign the discrete nanotube conducting states with the Fermi level of the superconductor, so each Josephson junction can be turned "on" or "off." In addition, the parts of the SQUID most sensitive to magnetic fields are the Josephson junctions, and the SWNT-SQUID would be able to detect the magnetization of a single molecule attached to one of the SWNT Josephson junctions (illustrated by the blue sphere in Fig. 4.28b).

Only a small number of devices and measurements have been presented, while there are a huge number of examples of hybrid nanostructures. Some more examples are presented in Chapter 5, Section 5.3. The important conclusion from this section, however, is that by combining top-down and bottom-up methods, it has become feasible to make electrical connections from our macroscopic world to individual nanoparticles and nanotubes. This paves the way to new generations of devices that can probe and study individual molecules.

4.4   IMAGING, PROBING, AND MANIPULATING NANOSTRUCTURES

4.4.1   Scanning Tunneling Microscope (STM)

The scanning tunneling microscope (STM) was first developed at the IBM laboratories in Zurich by Gerd Binnig and Heinrich Rohrer in 1981 and earned the inventors a share in the 1986 Nobel Prize in Physics. In fact the prize that year was awarded entirely for microscopy techniques and was split between Ernst Ruska for his work on electron optics and the invention of the electron microscope (see Section 4.4.6) and Binnig and Rohrer for the invention of the STM. The STM caused quite a stir in the 1980s showing beautiful images of surfaces with atomic resolution and has since become a major component of the nanotechnology toolkit, being able to not only image but also probe electronic states and manipulate atoms and molecules. It is at the heart of a suite of measurement techniques that is generally known as scanning probe microscopy (SPM).

Like many great inventions, the central idea is relatively simple. At the heart of any scanning probe microscope is a sharp tip that has been etched or processed so that the end has a radius of a few nanometers, and ideally there is a single atom protruding at the tip (Fig. 4.29a). Very fine movements of the tip are achieved by mounting it on a piezoelectric device. This is a ceramic that changes size when a voltage is applied to it, and it is possible to get controlled movements with sub-nanometer precision. The normal configuration, shown in Fig.4.29b, is to mount

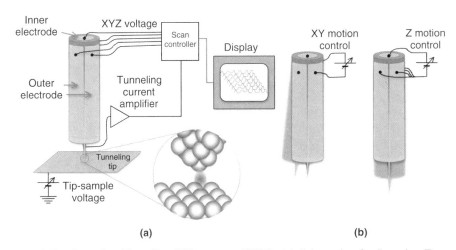

(a)   (b)

Fig. 4.29   Scanning Tunneling Microscopy (STM). (a) Schematic of a Scanning Tunneling Microscope with an atomically sharp tip brought to within $\sim$1 nm from a surface. A voltage between the tip and the (conducting) sample initiates a quantum mechanical tunneling current, which is sensitive enough to height to detect the atomic corrugation of the surface. The tip can be scanned at constant height and the image built up by plotting tunneling current on an XY map. Alternatively the control system can adjust Z to maintain a constant tunneling current and plotting the Z-control voltage on an XY map forms the image. The latter is the usual mode. (b) Schematic of a piezo-electric tube scanner. The outer electrode is in four quadrants and applying a voltage between opposite quadrants bends the tube in that direction. Applying a voltage between all four quadrants of the outer electrode and the inner electrode changes the length of the tube so full XYZ control (to sub-nm precision) is obtained.

the tip on a piezoelectric tube with four strip electrodes spaced evenly around the tube. Putting a voltage across opposite pairs bends the tube to produce a lateral motion of the tip; and a voltage applied between the inner and outer electrodes changes the height, giving full XYZ control. Using the tube scanner, the tip is brought to within about 1 nm of the surface, and a small voltage applied to the tip will initiate a current flow by quantum mechanical tunneling between the sample and the tip if the sample is conducting. Note that this current is classically forbidden and only flows because the quantum mechanical uncertainty principle provides some ambiguity about whether an electron is actually in the tip or the sample. See Advanced Reading Box 4.7 for a more formal description of the tunneling.

ADVANCED READING BOX 4.7—SIMPLE QUANTUM THEORY OF TUNNELING

The electronic states in an STM tip placed a microscopic distance away from a metal sample can be represented in one dimension by the very simple energy level diagram below. The Fermi level in the sample and tip are aligned and electrons can tunnel through the barrier of height V_0 from a filled state at the Fermi level in the tip to an empty state around the Fermi level in the sample. For the moment the position of zero energy (the floor of the potential well) is ambiguous, but we return to this below. The electrons at the Fermi level in the tip have a kinetic energy E relative to this floor, and they can be considered as incident on the barrier from the left.

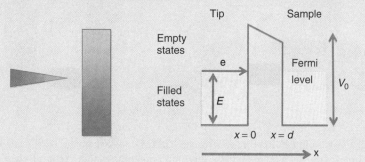

The wave function of the incident electrons in the tip is obtained in the usual manner by solving the 1D Schrödinger equation in the three regions $x < 0, 0 < x < d$ and $x > d$ of the barrier and matching the amplitude and gradient of the wavefunction at the boundaries. The 1D Schrödinger equation is

$$\frac{\hbar^2}{2m}\frac{\partial^2 \psi(x)}{\partial x^2} + (E - V)\psi(x) = 0 \qquad (4.5)$$

with

$$V = 0 \qquad \text{for } x < 0,$$
$$V = V_0 \qquad \text{for } 0 < x < d,$$
$$V = 0 \qquad \text{for } x > d.$$

Rewrite (4.5) as

$$\text{Regions 1\&3}: \frac{\partial^2 \psi(x)}{\partial x^2} + k^2 \psi(x) = 0 \text{ with } k^2 = \frac{2mE}{\hbar^2} \text{(regions 1 and 3)}$$

(4.6)

$$\text{and Region 2}: \frac{\partial^2 \psi(x)}{\partial x^2} + p^2 \psi(x) = 0 \text{ with } p^2 = \frac{2m(E - V_0)}{\hbar^2} \text{(p}^2 \text{is negative)}.$$

(4.7)

The problem is most easily solved if the solutions are written in the form

$$\psi(x) = \begin{cases} Ae^{ikx} + Be^{-ikx} \, (x < 0) \\ Ce^{px} + De^{-px} \, (0 \le x \le d) \\ AS(E)e^{ik(x-d)} \, (x > d) \end{cases}$$

(4.8)

Then the function $S(E)$ is the energy-dependent tunneling matrix element. Since we can choose an arbitrary amplitude (A) for the electron arriving at the barrier, we can set $A = 1$ so that the square of $S(E)$ is equal to the probability that an electron at $x = 0$ "finds itself" at $x = d$ moving away from the barrier (i.e., tunnels). Matching the amplitude and gradient of the wavefunction at $x = 0$ and $x = d$ gives

$$1 + B = C + D,$$
$$ik - ikB = pC - pD,$$
$$Ce^{pd} + De^{-pd} = S(E),$$
$$pCe^{pd} - pDe^{-pd} = ikS(E).$$

(4.9)

Solving these four simultaneous equations gives

$$S(E) = \frac{2ikp}{2ikp \cosh(pd) + (k^2 - p^2)\sinh(pd)}.$$

(4.10)

The transmission probability is

$$T(E) = |S(E)|^2,$$

(4.11)

which from (4.10) gives

$$T(E) = \left[1 + \frac{\sinh^2(pd)}{4(E/V_0)(1 - E/V_0)}\right]^{-1}. \tag{4.12}$$

For an electron De Broglie wavelength $\ll$ barrier width, d, that is, $kd \gg 1$ (converted to conventional units, this becomes $\sim 5.1d\,(nm)\sqrt{E}\,(eV) \gg 1$, which is easily satisfied), (4.12) becomes

$$T(E) \approx 16\frac{E}{V_0}\left(1 - \frac{E}{V_0}\right)\exp\left(-\frac{2d}{\hbar}\sqrt{2m(V_0 - E)}\right). \tag{4.13}$$

To get typical values for a metal, take the bottom of the conduction band as the energy zero. Then typically an electron at the Fermi level will have a kinetic energy, E, of 12 eV. The barrier height will be this plus the work function, which will typically be 4 eV so the total barrier height is $V_0 = 16$ eV. The transmission probability (proportional to the tunneling current) given by equation (4.13) is plotted below for these values of E and V_0 on a logarithmic scale. It is evident that a change in d of only 0.1 nm changes the tunneling current by more than a factor of 10, explaining the sensitivity of the STM to atomic corrugation. So far we have not taken account of the voltage applied between the sample and the tip, but this is required in order to align filled states at the Fermi level of the sample or the tip with empty states on the other side.

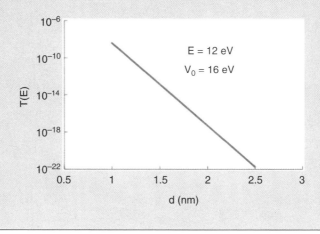

The key point is that the tunneling current is extremely sensitive to the separation between the tip and the surface and, given that the tip can be atomically sharp, is able to detect the atomic scale corrugation in the sample surface. There are two modes in which to operate the device. One is to sweep the tip at a constant height above the surface and plot the tunneling current as a function of X

and Y position of the tip (constant height mode). The other and more common mode is to get the control system to sweep X and Y and alter Z to maintain a constant tunnelling current (constant current mode). In this case the picture is formed by plotting the Z voltage of the tube scanner as a function of X and Y.

Some images demonstrating the atomic resolution of the STM are shown in Fig. 4.30. Figure 4.30a is an image of a Si surface cut along the (111) face showing the now well-known "7×7" structure. It is common for surfaces not to show the atomic structure of the ideal face—that is, assuming that the atoms

Fig. 4.30 Atomic resolution STM images. (a) Si(111) surface showing (7×7) reconstruction and some defects (missing atoms). Reproduced with permission from Omicron NanoTechnology GmbH (http://www.omicron.de). (b) Self-organized 6-atom In islands grown on the Si(111) 7×7 surface. Reprinted with permission from [50]. Copyright 2002 by the American Physical Society. (c) Pt(100) surface showing reconstruction. Reproduced with the permission of Elsevier Science from A. Borg et al. [51]. (d) Au (100) surface showing reconstruction. Reproduced from Wikipedia web page: http://en.wikipedia.org/wiki/Scanning_tunneling_microscope.

are frozen into their crystal lattice positions along the cut. The presence of the surface produces a reorganization of the electrons responsible for bonding; and often the lowest energy of the surface occurs when the atoms have reordered themselves into a new arrangement, a process known as reconstruction. The new structure is conventionally represented as a superstructure imposed on top of the ideal face. In the case of Si the superstructure repeats every 7×7 unit cells of the ideal crystal termination. As is often the case, the sample has to be heated in ultra-high vacuum for the reconstruction to take place. Note the realism of the image showing the significant number of defects, such as missing atoms. This is very common, and the invention of the STM showed how difficult it is to get a defect-free perfect surface.

Figure 4.30b shows a nice example of self-organization when a sub-monolayer quantity of In is deposited on top of the Si(111) 7×7 surface. The In atoms organize themselves into perfect triangular islands containing six atoms each, and the islands are ordered in a regular array. There are many examples of such self-organization when one material is grown on another and is a method for producing supported nanoparticle arrays on a surface.

Figures 4.30c and 4.30d show atomic resolution images of Pt and Au cut along the (100) crystal face. The atomic corrugation in metals is much less than that of semiconductor surfaces like Si, and obtaining such high-resolution pictures is not easy. Both surfaces show reconstructions, which were known about prior to the development of the STM from electron diffraction measurements. These showed hugely complicated diffraction patterns with superstructures that were very large relative to the underlying lattices. With the clear atomic resolution images that emerged with the availability of STMs, the underlying simplicity of the reconstruction emerges and it is evident, in both cases, that it corresponds to a "rumpling" of the surface.

It is very easy to get carried away with the beautiful and clear atomic resolution images such as those in Fig. 4.30 and think of them as "seeing atoms." When considering the process by which they are obtained, however, great care has to be taken in interpreting the images. For example, the process involves tunneling of electrons in filled states in the sharp tip to empty electronic states in the surface, or vice versa. Locally, where these states are in relation to the position of the atoms (as, say, defined by the position of the atomic nucleus) is not clear. For example, changing the bias on the tip when scanning a Si surface apparently "moves" the atoms as tunneling into different states occurs. Another example is the graphite surface on which only half the atoms that should be there appear in the STM image. This is due to a slight unevenness of the surface so that every second atom is depressed slightly. One other problem with the STM is that it needs a conducting surface to get an image. Despite these problems, STM has revolutionized the study of surfaces and, as we will see below, it is not restricted to imaging but can probe the local electronic structure and also manipulate structures on the surface.

Since the 1980s there have been steady improvements in STM performance as a result of technical developments in vibration isolation, control of the piezoelectric scanners, and so on. In addition, sample environments have become much more sophisticated, allowing surfaces to be scanned in ultra-high vacuum (UHV), which is vital for the surface preparation of most materials. For example, all the images shown in Fig. 4.30 were obtained in UHV. Scans can now also be performed with the sample maintained at temperatures from millikelvin to over 1300 K. One thing that has not changed is that the most important component for high-resolution performance is the tunneling tip, which remains a somewhat hit-and-miss affair when attempting to reach ultimate resolution. They can be produced in a variety of ways, and the normal method is to use electrochemical etching of a tungsten wire to produce tips like the example shown in Fig. 4.31a. The inset SEM image has a 650,000× magnification, close to the resolution limit of an SEM and it is clear that the radius at the end of the tip is <10 nm.

Recently the STM, central to much of nanotechnology, has itself benefited from nanotechnology, and ultra-sharp tunneling tips have been produced by attaching carbon nanotubes to conventionally synthesized tips. An example produced by Chin and co-workers at the institute of Physics in Taiwan [52] is shown in Fig. 4.31b. In this process a multi-walled carbon nanotube fixed to a movable stage is attached to a conventionally produced tip (Au in this case) and secured by the "welding" process using the TEM electron beam described in Section 3.12, Chapter 3. The tube is broken roughly in the middle by passing a current through it. Since the tube burns off layer by layer, the resulting morphology around the break is the "telescope" structure observed in the high-magnification picture. Thus the tip has the stiffness of a MWNT but with an end the size of a SWNT. Clearly, the tip radius is well below 5 nm.

4.4.2 Manipulating Atoms and Molecules with STM

Within a decade of the invention of the STM, it was discovered that the tip could also be used to move and position atoms and molecules on a surface. This was first achieved by Don Eigler and Erhard Shweizer at the IBM Almaden Research Laboratory in 1990 [53] and marked a milestone in nanotechnology. In some ways it marked the final realization of the dream originally articulated by Richard Feynman in 1959 at a speech to the American Physical Society at the California Institute of Technology (Caltech). This speech is generally regarded as the original visionary definition of nanotechnology, and in it he stated:

> ... I am not afraid to consider the final question as to whether, ultimately—in the great future—we can arrange the atoms the way we want; the very atoms, all the way down! What would happen if we could arrange the atoms one by one the way we want them (within reason, of course; you can't put them so that they are chemically unstable, for example). ...

The original demonstration of this ability by Eigler and Shweizer was to write "IBM" in xenon atoms on a Ni surface as illustrated in Fig. 4.32. In this experiment the Ni(110) surface was first cleaned *in situ* in UHV to remove contaminants

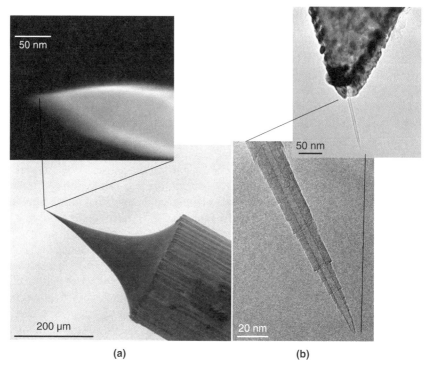

(a)   (b)

Fig. 4.31 STM tunneling tips. (a) SEM image of an STM tip produced by electro-chemical etching of a tungsten wire. The highly magnified inset of the end of the tip, at a magnification of 650,000× and close to the resolution limit of an SEM (hence the slightly blurred image), demonstrates that the process can produce a radius at the end <10 nm. Reproduced with permission from Dr. Joanna Millunchick, Department of Materials Science and Engineering, University of Michigan. (b) TEM images of a conventionally processed STM tip with an attached multi-walled carbon nanotube. This is heated inside the TEM to burn off outer layers and produce the 'telescope' structure shown in the high magnification image. The tip thus has the stiffness of a MWNT but the end has the size of a SWNT with a tip radius <5 nm. Reproduced with the permission of the Institute of Physics from S.-C. Chin et al. [52].

and cooled to 4 K. It was then dosed with a sub-monolayer quantity of xenon atoms, which adsorbed at random positions on the surface (Fig. 4.32a) attracted to the surface by the van der Waals force (see Chapter 3, Advanced Reading Box 3.4). Using the mechanism described below, individual atoms were chosen and moved to new positions shown in Figs. 4.32b and 4.32c.

The method used to move atoms is illustrated in Fig. 4.33. First the STM tip was scanned in constant current mode (Fig. 4.33a) using a relatively low tunneling current (10^{-9} A) in order not to move the atoms, which are weakly bound to the surface, during scanning. Each atom caused the STM tip to rise by 0.16 nm in order to maintain a constant current. Next the tip was placed over

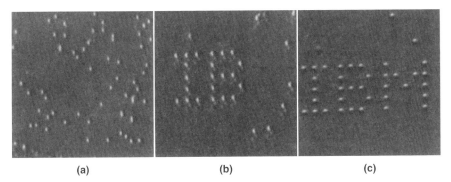

(a)   (b)   (c)

Fig. 4.32   First demonstration of manipulating individual atoms using an STM.
Controlled positioning of individual xenon atoms on a Ni surface using an STM first
demonstrated by Eigler and Shweizer in 1990 [53]. Reproduced with the permission of
the Nature Publishing Group from D. M. Eigler and E. K. Shweizer [53].

a chosen atom in the image, and the set current was increased by at least an
order of magnitude. The STM control system moves the tip closer to the surface
to maintain the set current (Fig. 4.33b); and as discussed in Advanced Reading
Box 4.7 , an order of magnitude increase in current requires a movement toward
the surface of ∼0.1 nm. If the current is chosen correctly, the van der Waals
force between the tip and the atom will increase to a level that is not sufficient
to pull the atom off the surface but high enough to keep the atom under the tip
as it is moved. The tip was then moved (Fig. 4.33c) with the control system
maintaining the high tunneling current (i.e., the low height above the surface).
Finally the current was reset back to the low imaging value and the new position
was imaged (Fig. 4.33d).

The system chosen—that is, xenon atoms on Ni—was particularly amenable to
this experiment due to the relatively large size of the xenon atoms and their weak
bonding to the substrate, but it passed the important milestone and demonstrated
that atomic manipulation is possible. Since 1990 the art has gradually improved,
and a variety of atoms and molecules have been individually positioned on various surfaces using STMs. Figure 4.34 shows a selection of "quantum corrals"
assembled using Fe atoms on a Cu(111) surface at the IBM Almaden laboratory [54]. This level of sophistication in the manipulation takes the field from
demonstrating proof of principle to active research in the quantum mechanics
of electron waves in metal surfaces within well-defined boundaries. Waves are
clearly seen within the corals, and these are due to Cu electrons in the metal
surface scattering from the boundaries and interfering. The resulting charge density wave is observed by the STM since the tunneling current is sensitive to the
charge density on the surface. Single-atom manipulation has come a long way
since 1990; and new abilities have been demonstrated, for example, pulling atoms
on to the tip and re-depositing them in a different place [55]—so-called vertical

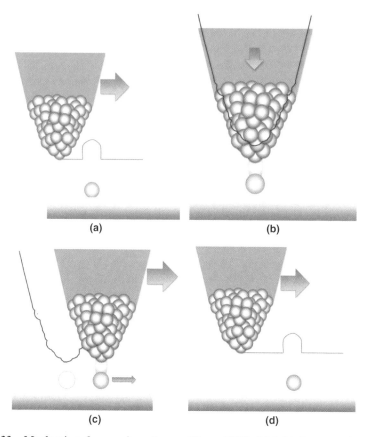

Fig. 4.33   Mechanism for moving atoms with an STM. (a) Initially a scan is taken in constant current mode at a sufficiently low current that the scan only images and does not move atoms (10^{-9} A). An adatom causes a change in height (0.16 nm for Xe) at its position. (b) The tip is moved over the atom and moved closer to the surface by increasing the tunneling set current to $> 10^{-8}$ A. The tip-atom interaction is increased but not sufficiently to lift it from the surface. (c) The tip is moved, maintaining the higher tunneling current, to the desired position. (d) The tunneling current is reset to its low imaging value and the atom is imaged in its new position.

manipulation. The precision demonstrated in Figs. 4.32, however, has not been achieved at room temperature, which remains an important future milestone.

Although not yet achieved with atoms, room temperature manipulation of C_{60} fullerenes on surfaces has been demonstrated by Cuberes et al. [56] at the IBM Zurich laboratories, where they assembled the now iconic C_{60} abacus on a Cu surface (Fig. 4.35). This experiment utilized a Cu crystal with a regular array of atomic steps along the surface, and the fullerenes were assembled in lines along the steps. The mechanism of moving fullerenes is different from atoms, which "slide" across the surface under the tip. In the case of C_{60}, it was

Fig. 4.34 Quantum corrals assembled from Fe atoms on a Cu(111) surface.
Different-shaped enclosures made from individual Fe atoms manipulated by STM on
a Cu(111) surface at 4 K. Note the waves within the corrals produced by scattering and
interference of the electrons in the Cu(111) surface and the different patterns arising from
the different shapes of barriers. Reproduced with the permission of Elsevier Science from
M.F. Crommie et al. [54].

shown by Moriarty et al. [57] at the University of Nottingham that the molecules
roll across the surface, much like a football. This was evidenced by resolving
intramolecular features with the STM and showing that they changed with the
position of the molecule on the surface. Fujiki et al. at Okayama University
demonstrated controlled removal of individual C_{60} molecules from a close-packed
layer adsorbed on a surface, thus demonstrating the possibility of drawing an
image with 1-nm pixels [58].

4.4.3 Scanning Tunneling Spectroscopy (STS)

A further ability of an STM is its ability to probe local electron states with a
spatial resolution $\sim$0.5 nm using scanning tunneling spectroscopy (STS). The
initial development of STS dates back to Binnig and Rohrer, who noticed from
the earliest images that the appearance of atoms in a Si(111) 7×7 surface was
altered by using different tip-sample bias voltages. It has since developed into a
powerful spectroscopy tool and is incorporated into all commercial STMs. The

Fig. 4.35 C$_{60}$ abacus. C$_{60}$ molecules manipulated using an STM on a stepped Cu surface into lines along the steps in the form of an abacus. The structure was built at room temperature. Reproduced with the permission of the American Institute of Physics from M. T. Cuberes et al. [56].

experiment is relatively straightforward and involves placing the tip at a desired position over a molecule or an atom and recording the tunneling current as a function of tip-sample bias voltage.

As the tip-sample bias is changed, different states are accessible to the tunneling electrons; thus measuring the current, I, as a function of the bias voltage, V, provides information on the electron states in the sample at the position of the tip. More specifically, the gradient of the $I-V$ curve, the so-called tunneling conductance, is, within certain approximations, proportional to the local density of electron states or LDOS of the sample. In a metal the density of conduction band electron states varies relatively smoothly with energy, so STS from a clean metal surface shows rather featureless spectra while a molecule sitting on the surface will exhibit sharp features in STS corresponding to specific energy levels. Advanced Reading Box 4.8 gives a more detailed description of the information obtained with STS.

An illustrative example of information available using STS is shown in Fig. 4.36. These data were measured from C$_{60}$ molecules on a Si(100) 2×1 surface by the Optical Sciences group at the University of Arizona, who also obtained the nice STM image shown in Fig. 3.12, which is reproduced in the inset in Fig. 4.36. As discussed in Chapter 3, Section 3.6, C$_{60}$ has an absence of electronic states over $\sim$2 eV in energy known as the HOMO–LUMO (highest occupied molecular orbital–lowest unoccupied molecular orbital) gap. This

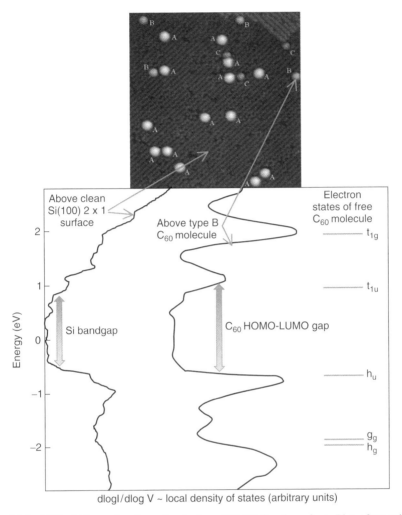

Fig. 4.36   STS of C$_{60}$ molecules adsorbed on Si(100) 2 × 1 surface. Plot of tunneling conductance, dlog$I$/dlog$V$ (the logarithm of I and V has been used to increase sensitivity), which is related to the sample LDOS taken from a position above the clean Si(100) 2 × 1 surface and from a position above a type B C$_{60}$ molecule. The bandgap in the Si surface and the HOMO-LUMO gap in the C$_{60}$ electron states are evident. The column on the right indicates the calculated electron states of a free C$_{60}$ molecule. Reproduced with the permission of Elsevier Science from X. Yao et al. [Ref. 12], Chapter 3.

gives it an electronic structure reminiscent of a semiconductor, whose electronic behavior is characterized by an energy gap between its valence electrons and conduction electrons. In STS the gap should reveal itself as a significant reduction in the tunneling conductance as the voltage is swept through the appropriate values. The presence of the bandgap in the Si(100) 2 × 1 surface is evident in the STS spectrum taken from a position above the clean surface. Moving the tip above a type B C_{60} molecule (see Chapter 3, Section 3.6 for a discussion of the different types of adsorbed C_{60}), clearly shows the HOMO–LUMO gap of the fullerene.

ADVANCED READING BOX 4.8—INFORMATION OBTAINED FROM SCANNING TUNNELING SPECTROSCOPY (STS)

The density of states function $g(E)\,dE$ of a material is the number of states per unit energy in the energy range E to $E + dE$ and is a measure of how densely packed in energy the states are. STS probes the local density of states function, $g_L(E)$, which can vary with position on a sample, with a spatial resolution of ~0.5 nm. The figure shows the local density of states (LDOS) of the tip (a metal) and that of the sample, which is assumed to be a metal substrate with an adsorbed molecule at the position of the tip. The LDOS of the tip and the substrate are smooth functions ($g(E) \propto \sqrt{E}$ for a simple metal) irrespective of position, while the molecule has sharp molecular orbitals for the electrons to tunnel into or out of. With no tip-sample bias the Fermi levels of the tip and substrate are aligned and the energy of the molecular orbitals are fixed with respect to the Fermi level of the substrate. Applying a tip-sample bias separates the Fermi levels by the bias voltage (a negative voltage on the tip depresses the sample Fermi level as shown), and all the empty states in the sample that are aligned with filled states in the tip are available for tunneling. These are indicated by the shaded region in the sample LDOS.

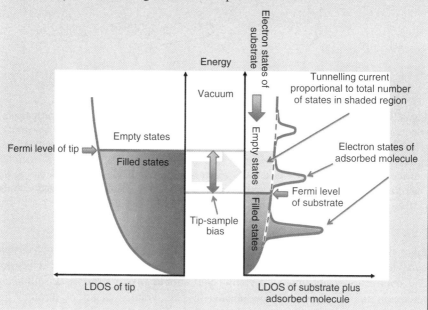

Making the gross assumption that the LDOS in the tip is constant, the tunneling current, I, is proportional to the Integral of all the electronic states in the sample between the Fermi level and the bias voltage. Thus, for a bias voltage V we have

$$I \propto \int_0^{eV} g_{L,\text{sample}}(E + E_F)\, dE.$$

Differentiating the current with respect to the bias voltage, we get

$$\left. \frac{dI}{dV} \right|_V \propto g_{L,\text{sample}}(E + E_F).$$

Thus within the approximation of a constant $g_L(E)$ of the tip, the sample LDOS is proportional to the gradient of the measured $I-V$ curve. If we want to get an accurate LDOS of the sample, we need to know the LDOS of the tip at the same position; but to detect the position of sharp molecular orbitals or to measure HOMO–LUMO gaps as in Fig. 4.36, the above equation is adequate.

4.4.4 Atomic Force Microscopy (AFM)

A problem with the STM is that it is unable to obtain images from insulating surfaces, and this led to the development of the atomic force microscope (AFM) by Binnig et al. in 1986 [59]. It is the cousin of the STM in that at the heart of it is the same ultra-sharp tip found in an STM except that in the AFM the tip is fabricated on the end of a tiny cantilever as shown in Fig. 4.37a. As in the STM, piezoelectric control is used to position the cantilever and tip in x, y, and z with a precision of ~ 0.1 nm. When it is mounted in the microscope, a laser is reflected from the back of the cantilever onto a position-sensitive detector. There is a sufficiently long optical lever effect that cantilever deflections ~ 0.1 nm can be detected. As the tip is brought toward a sample, at separations below ~ 50 nm there is initially an attractive force (the van der Waals force; see Advanced Reading Box 3.4 , Chapter 3) that bends the cantilever a detectable amount toward the surface. The deflection increases for a while; but below a critical separation, the force becomes repulsive and the cantilever deflects upwards. This corresponds to the tip touching the surface.

The sequence of events as a cantilever approaches the surface is shown in more detail in Fig. 4.38. The measured force as a function of height is depicted by the blue curve for approach and the red curve for withdrawal, with the curves shifted slightly for clarity. The images along the top depict the state of the cantilever at various heights. A long way from the surface (Fig. 4.38a), there is no force and the cantilever is relaxed. On approaching the surface, the van der Waals attraction between the tip and the surface produces a measurable downward deflection of the

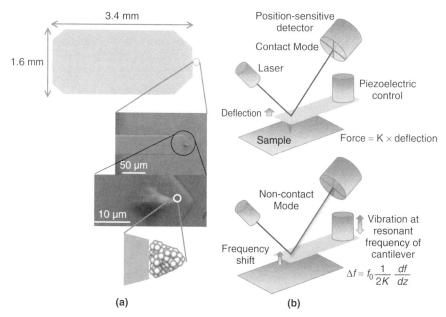

Fig. 4.37 **Atomic Force Microscopy (AFM).** (a) A standard commercial cantilever is fabricated on a Si chip with a thickness of ~300 μm and the lateral dimensions shown. This acts as a sturdy platform for mounting the cantilever in the microscope. The cantilever is typically 100–300 μm long depending on the application with the tip fabricated on the end. (b) To measure the cantilever deflection a laser is reflected from the back of the cantilever into a position–sensitive deflector. The long optical lever arm enables deflections of <0.1 nm to be detected. The two scanning modes normally used are shown in the top and bottom figures. In contact mode (top), the deflection of the cantilever is measured as the tip is scanned across the surface and is plotted as a function of x and y to obtain the image. In non-contact mode (bottom) the cantilever is vibrated at its resonant frequency (usually ~10-100 kHz) and the frequency shift (proportional to the force gradient) is plotted as a function of x and y to obtain the image.

cantilever (Fig. 4.38b). At some critical distance the random thermal vibrations of the cantilever cause a force runaway; that is, on a particularly large swing the tip experiences a larger force that causes it not to return to its equilibrium position and on the next swing it finds itself even closer to the surface. This makes the tip "snap-on" to contact with the surface (Fig. 4.38c), producing a maximum downward deflection of the cantilever. As the approach continues, the cantilever is bent straight again to its relaxed zero-force position (Fig. 4.38d). Moving past this point causes the cantilever to bend upwards (Fig. 4.38e). Withdrawal from the surface produces a similar force curve; but since the tip is already in contact with the surface, the cantilever has to be withdrawn past the "snap-on position" to get it to "snap-off."

There are two basic modes in which the instrument is used to obtain an image of a surface depicted as contact and non-contact in Fig. 4.37. In contact mode

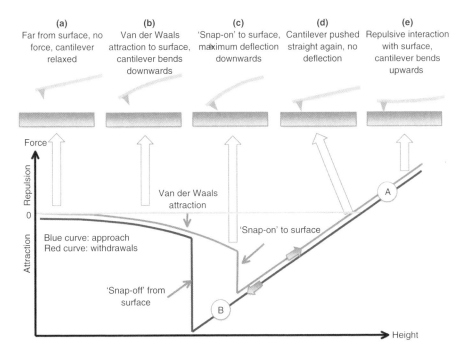

Fig. 4.38 **Deflection of cantilever approaching surface.** Schematic of cantilever deflection as a tip approached a surface. The measured force as a function of height is depicted by the blue curve for approach and the red curve for withdrawal with the curves shifted slightly for clarity. (a) Far from the surface there is no force and the cantilever is relaxed. (b) Approaching the surface the Van der Waals attraction starts to produce a measurable deflection downwards. (c) At some critical separation the thermal vibration of the cantilever will cause a force runaway and the tip will 'snap' into contact with the surface. At this point there will be a maximum measured downward deflection. (d) As the cantilever is pushed further down it is straightened back to the relaxed zero force position. (e) Approaching further the cantilever is bent upwards. Withdrawal follows a similar path except that the cantilever has to be withdrawn further than the 'snap-on' position before it 'snaps-off'. Point A is used as the set-point for positive (repulsive) force scanning and point B for negative (attractive) force scanning.

the tip is approached toward the surface until "snap-on" and then driven past this point to point A (positive or repulsive force) or withdrawn to point B (negative or attractive force). It can then be scanned in x and y with the control system moving the cantilever in z to maintain a constant cantilever deflection (force). The z-piezo voltage is plotted as a function of x and y to obtain the image. This is equivalent to the constant current mode of the STM and is the normal way of imaging in contact mode. Just as the STM can have a positive or negative set-point current, the AFM can be scanned with a positive (point A) or negative (point B) set-point force. With small set-point forces (usually on the order of nanonewtons) the touch of the AFM is particularly gentle and does not damage surfaces, even at the atomic level. Often, however, adsorbed atoms and molecules are moved by the tip, and this is one of the problems circumvented by using the non-contact mode.

In non-contact mode the tip (as the name implies) never touches the surface. The z-piezo voltage is modulated to make the tip vibrate at its resonant frequency, f_0 (usually in the range 10–100 kHz), and the vibration of the end of the cantilever is detected by the oscillating signal in the position-sensitive detector. As the tip approaches the surface, the attractive force causes the frequency to shift by an amount, Δf, given by the equation in Fig. 4.37, where K is the stiffness constant of the cantilever. It is evident that the frequency shift is related to the gradient of the force with height, but this gets larger as the surface is approached like the force itself. To obtain an image, the tip is scanned in x and y with the control system moving the cantilever in z to maintain a constant frequency shift, Δf. The image is a map of the z-piezo voltage as a function of x and y.

In either mode it is possible to scan at constant height and plot the force or the frequency shift as a function of x and y to obtain an image, but this is not so common in topographical imaging. It is, however, used in other AFM-based measurements such as magnetic force microscopy (see below). Just as with the STM, the very high spatial resolution of the AFM is given by the ultra-sharp tip that can end at a single atom and is able to detect the atomic corrugation of a surface. The instrument is also benefiting from the same developments in tip technology such as the carbon nanotube tip (Fig. 4.31). There is a general folklore that the spatial resolution of AFM is not quite as good as STM, but for fundamental reasons it can be even better. The STM utilizes tunneling electrons into states around the Fermi level, which tend to be spatially extended, whereas AFM relies on the force between atoms, which is generated by all the electrons including the highly localized core electrons around the nucleus. For non-contact measurements the cantilevers used are shorter and stiffer than those specialized to contact measurements so that they have a high resonant frequency and a smaller amplitude of oscillation. Commercial Si cantilevers for non-contact imaging generally have a stiffness constant of about 10 N/m, and the amplitude of oscillation at the tip when they are at resonance is typically about 10 nm. Recently new high-stiffness cantilevers have been fabricated in quartz with stiffness constants >1000 N/m and an oscillation amplitude of <1 nm [60]. Figure 4.39 shows an AFM image taken in non-contact mode of a Si(111) 7 × 7 surface using a cantilever with stiffness constant of 1800 N/m and demonstrates atomic resolution [61].

Since the initial development of the AFM in 1986, it has proved enormously versatile and has been used to measure not only topography but also magnetic forces, electrical forces, and frictional forces between a tip and a sample. It has provided data from environments varying from samples submerged in liquids to ultra-high vacuum. It has also provided data on a strange force known as the Casimir force resulting from the quantum fluctuations of vacuum (see Chapter 8). A review of all the uses of AFM is beyond the scope of this book, but below is presented what is probably the second most common usage after topography

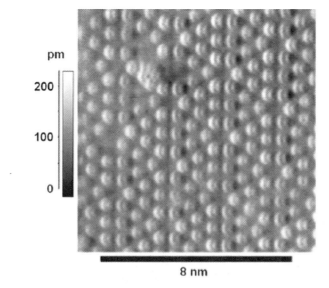

pm

200

100

0

8 nm

Fig. 4.39   Atomic resolution non-contact AFM image of Si(111) 7 × 7 surface. Non-contact AFM image taken in ultra-high vacuum of a Si(111) 7 × 7 surface using a high-stiffness cantilever ($K = 1800$ N/m). The resonant frequency was 16.86 kHz. Reproduced with the permission of the American Association for the Advancement of Science (AAAS) from F. J. Giessibl et al. [61].

measurements—that is, measurements of the magnetic force between a magnetic tip and a magnetic sample.

Magnetic force microscopy (MFM) uses magnetic tips that are synthesized by coating a conventional tip with a thin magnetic film as illustrated in Fig. 4.40a. One problem in obtaining the magnetic force is separating it from the much larger interatomic forces as a tip is scanned over a surface. A solution patented by Digital Instruments called "lift mode" is illustrated in Fig. 4.40a. For each raster of a two-dimensional scan, the tip is first used to obtain a conventional non-contact AFM topography image (scan 1). The tip is then lifted a specified distance from the surface (typically 50–100 nm), and the scan is repeated with the control system moving the tip along the previously recorded topography (scan 2). Any remaining variations in force across the surface are due to the long-range magnetic force between the tip and the sample. An MFM scan in lift mode thus records both the topography and the magnetic force at the same time. Figure 40.40b shows an MFM scan of the magnetic film on the surface of a hard disk used for storing digital data. The topography image on the left simply shows a rough surface; but using lift mode, the image on the right shows the magnetic pattern of the stored data in two tracks. With MFM, since a magnetic tip is brought close to a magnetic surface, care must be taken not to remagnetize the tip or the surface during imaging. The possibility of using the tip to modify the

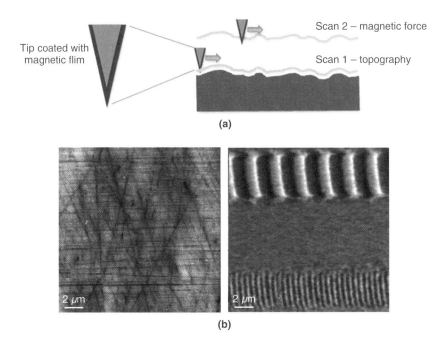

(a)

(b)

Fig. 4.40   Magnetic Force Microscopy (MFM) of a magnetic sample using lift mode.
(a) Magnetic Force Microscopy (MFM) of a magnetic sample uses a tip coated with a magnetic material that experiences a magnetic force when it approaches the sample. To separate the topography of the surface from the variations in magnetic force a common approach is to use the 'lift mode' originally developed by Digital Instruments. With this method, for each line of a scan, the tip is scanned in non-contact mode at a height suitable to obtain an AFM topography image. The tip is then lifted a set height above the surface (~50–100 nm) and then scanned again following the line of the topography. Any remaining variation in the force is due to the magnetic interaction between the sample and the tip. (b) Simultaneous AFM topography and MFM image of a data storage disk showing the pattern of the stored data in the MFM image. Reproduced with permission from Nanosensors™, Switzerland (http://www.nanosensors.com).

magnetization on the surface can, however, be exploited, and in Chapter 5 the utilization of MFM to "write" data onto individual nanoparticles is discussed.

4.4.5   Dip-Pen Nanolithography (DPN)

Since its invention, the AFM has also been used as a tool for nanofabrication on surfaces by moving or modifying structures, but a particularly versatile method that has emerged in the last decade is known as dip-pen nanolithography (DPN). The development of the technique started with the realization that when obtaining AFM images in ambient air, water condenses around the tip and a meniscus is formed between the tip and the substrate. This was generally regarded as a problem because it inhibits the achievement of atomic resolution and normally steps are taken to carry out experiments in very dry air or, preferably, vacuum. In 1997 Piner and Mirkin [62] showed that there is a dynamic process in which water is either transported from the tip to the surface or vice versa, depending on the relative humidity and wetting properties of the surface. By moving the tip across a suitable substrate, they demonstrated that stable water layers could be deposited. Two years later the same team coated a tip with alkanethiol molecules and then set up suitable conditions of humidity for the molecules to be transported in the water meniscus from the tip to a gold surface. By then moving the tip across the surface it was possible to "write" lines of alkanethiols only 30 nm across as shown schematically in Fig. 4.41a. The name was coined because of

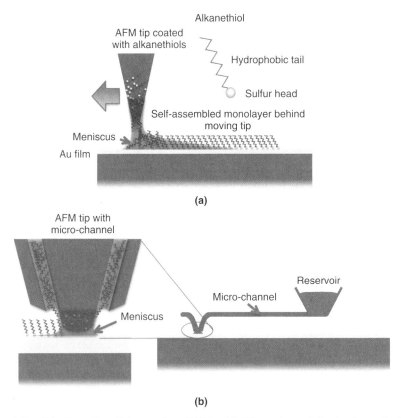

(a)

(b)

Fig. 4.41 Dip-Pen Nanolithography (DPN). (a) Illustration of the basic method of DPN. When an AFM tip is in close proximity to a surface in ambient air, water vapor condenses around the tip and forms a meniscus. If the tip is pre-loaded with a soluble molecule, under the right conditions the water transports the molecules onto the surface at the position of the tip making it possible to 'write' lines ~30 nm across in a manner similar to a dip-pen operating at the nanoscale. The figure shows the tip writing a line of alkanethiols, which form a self-assembled monolayer (SAM) within the written line. (b) More recent developments include a reservoir and a micro-channel that feeds the tip with ink.

the similarity to a dip-pen, where in this case the nib is the AFM tip, the ink is the alkanethiol-loaded water, and the Au surface is the paper. Following deposition, the written structures can be imaged using the same tip by either lowering the humidity or scanning at a much faster rate to prevent further deposition.

The alkanethiols were chosen for the initial demonstration because they condense into ordered arrays known as self-assembled monolayers (SAMs) as a result of their chemical structure. They have a sulfur head, which shows a special affinity for gold substrates, attached to a hydrophobic hydrocarbon chain. When they are in a solution on a gold surface, the sulfurs attach to the substrate and come together into an ordered crystalline structure while the hydrophobic tails point away from the metal surface. Thus a novel aspect of this nanofabrication technology is that it combines both top-down and bottom-up synthesis since the write process with the AFM tip belongs to the top-down category while the self-assembly of the molecules within the written lines and features is a bottom-up process.

Since the initial demonstration, the method has developed further and has included inks that are solid until heated so that deposition can be initiated by heating the tip and prevented by cooling it again. This makes it simple to switch between writing and imaging. Also, cantilevers and tips have been fabricated with an ink reservoir and a micro-channel feeding the ink to the AFM tip as illustrated in Fig. 4.41b. Dip-pen nanolithography has thus evolved into a fountain pen nanolithography tool. In addition, a range of nanostructures and molecules has been incorporated into the ink, including carbon nanotubes and antibodies [63] (see Chapter 6, Section 6.1.5).

One of the main technological applications of DPN is the synthesis of arrays of active biological molecules for the ultrasensitive detection of antigens or viruses. An example of the kind of detector array that can be produced is shown in Fig. 4.42. To begin with, the AFM tip is used to produce an array of dots inked with a chemical that will bind antibodies. A suitable substance is mercaptohexadecanoic acid (MHA), whose chemical formula is $HS(CH_2)_{15}CO_2H$. Just as with the alkanethiols, the sulfur atom bonds strongly to a gold surface and the MHA molecules form a SAM within the inked features. An important property of MHA is that it binds strongly to antibodies so an array of MHA dots such as the one illustrated in Fig. 4.42a exposed to a solution of antibodies will produce an antibody array whose active ends are exposed to the environment as indicated in Fig. 4.42b. As described in detail in Chapter 6, Section 6.1.5, antibodies are "Y"-shaped biological molecules in which the ends of the arms of the "Y" (the paratopes) are interchangeable so that they will bind to specific proteins, very much like a lock and key. Thus the antibodies chosen will be those whose paratopes are sensitized to the specific biological molecule to be detected.

The array is exposed to the suspension to be tested; and if any of the target molecules are present, they will attach to the paratopes of the antibodies. A successful targeting event is identified by the change in height of an AFM scan at that position in the array (Fig. 4.42c). The detection works at the single-molecule

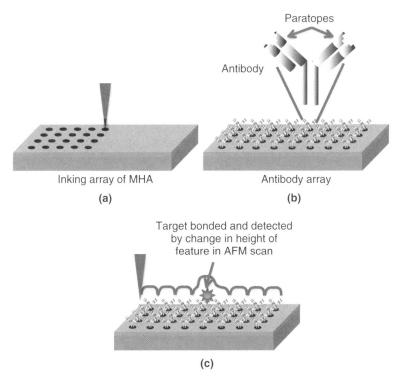

Fig. 4.42 Nanoarrays for the ultrasensitive detection of biological molecules. (a) First DPN is used to ink a regular array of nanoscale spots of a chemical that can bind a biologically active molecule. For example MHA will bind antibodies. (b) Antibodies are biological recognition agents whose 'paratopes' fit specific proteins by a 'lock and key' mechanism. Immersing the inked MHA array into a solution containing antibodies of the desired type will generate an array of antibodies whose paratopes are exposed to the environment. (c) After exposing the array to the solution to be tested, if any of the target molecules are attached to antibodies they can be detected by a change in height of the array element at that position in an AFM scan.

level; and the sensitivity is orders of magnitude higher than any other detection method, though it is limited in its working concentration range since it saturates with a relatively small number of detected molecules. Screening for the HIV virus has been demonstrated using the method [64]. It is clear that DPN will be an important tool for medical diagnosis because it is a powerful generic method for synthesizing arrays of a range of biological molecules including DNA, peptides, and proteins [63].

4.4.6 Electron Microscopy

Throughout this chapter, images of nanostructures obtained using transmission electron microscopes (TEMs) and scanning electron microscopes (SEMs) have been displayed. This section provides a very brief description of the operation of these two types of electron microscope. A full review of transmission electron microscopy and its utilization for materials science is given in reference [65]. Like all fundamental particles, electrons behave as waves with a wavelength given by the De Broglie formula (see Advanced Reading Box 4.9). It is the very short wavelength of electrons at the typical energies used in electron microscopes that enables them to image tiny structures. Any probe that has a wavelength has an ultimate limit to the resolution it can provide, even if the optics in the microscope were perfect. This so-called *diffraction limit* of the resolution, for a given geometry of microscope is directly proportional to the wavelength of the probe particles. In an optical microscope the probe particles are photons with a wavelength around 500 nm while electrons in an electron microscope have a wavelength in the range 0.001–0.02 nm, depending on their energy. Glass optics used in microscopes and telescopes, having been developed for centuries, has reached a very high degree of perfection; and optical microscopes can image features right down to the diffraction limit, which in practice is ~1 μm, enabling them to image individual biological cells. In electron microscopes the wavelength of electrons is much smaller than an atom; but the optics (magnetic lenses) suffer from much greater aberrations than do their optical counterparts, so no electron microscope is able to image features with sizes at the diffraction limit. Modern aberration-corrected instruments are, however, achieving resolutions <0.1 nm, giving them true atomic resolution as demonstrated by the superb TEM image of a single Au nanoparticle shown in Fig. 4.43 obtained at the Nanoscience Centre at the University of York, United Kingdom.

Transmission electron microscopes use magnetic lenses to focus the electron beam and a rudimentary design is shown in Fig. 4.44b. It consists of a conducting coil within a soft Fe container that has a gap around the inner surface. The magnetic field produced by a current through the coil is zero around the container but emerges from the gap. Electrons moving through a magnetic field with a constant velocity experience a sideways force (Lorentz force) that is proportional to the magnitude of the magnetic field perpendicular to the direction of motion. With the geometry of magnetic lens shown, the Lorentz force increases away from the axis, so electrons moving furthest away from the axis experience the greatest deflection just as with an optical lens, though an electron's path through the magnetic lens is a spiral. Lenses of whatever type produce magnification by bringing together rays from the same point on an object to a single equivalent point at an enlarged image as illustrated in Fig. 4.44c. With a magnetic lens the image is rotated relative to the object; but apart from this detail, the basic optical design in a TEM is fundamentally the same as in a light microscope, with the important difference that magnetic lenses can be switched on or off or have their focal length changed by a simple adjustment of the current. In a real instrument there will

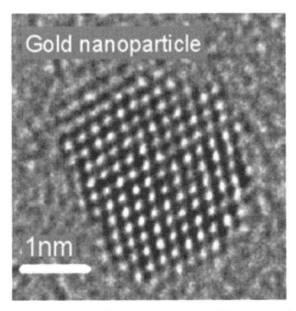

Fig. 4.43   High-resolution TEM image of Au nanoparticle. Image of an isolated Au nanoparticle obtained with an aberration-corrected TEM showing individual Au atoms. Reproduced with permission from Prof. Pratibha Gai at the Nanoscience Centre, University of York, UK.

be a number of magnetic lenses to (a) condition the electron beam before it passes through the sample and (b) focus and magnify the image after. There are also electrostatic deflectors to shift and raster the beam. A commercial TEM can be run in a number of different modes to obtain images or diffraction patterns.

The normal mode for imaging is to illuminate the sample with electrons from the back and the transmitted electrons are used to form the image. This means that both the sample and its support have to be transparent to electrons at the energies used in the TEM. A standard TEM sample holder used on all commercial instruments is shown in Fig. 4.45a and consists of a 3-mm grid available in a range of materials and grid sizes. This is then coated with the support for the sample, and a standard coating is a 100-nm-thick film of carbon. While this is transparent to electrons, scattering within the film does degrade the image; and for high-quality imaging of nanoparticles, a good alternative to a continuous film is a "lacy carbon" coating illustrated in the figure. This is mostly empty space; and if a nanoparticle sample is deposited on the grid, there is a good chance of particles sticking to the edge of one of the holes and suspended over a void as in the example shown in Fig. 4.45b for a collection of Fe nanoparticles.

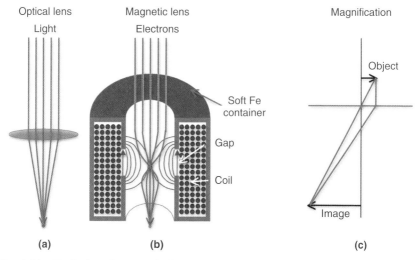

Fig. 4.44 Optical and magnetic lenses. (a) Optical lens. (b) Magnetic lens produced by a conducting coil within a soft Fe container with a gap on its inner surface. The magnetic field emerges only from the gap and is configured so that the Lorentz force on the electrons increases with increasing distance from the axis to produce the lensing effect. The electron path in the lens is a spiral and the image is rotated relative to the object. (c) A lens produces magnification by focusing all the rays from a given point on the object to a single point on an expanded image plane.

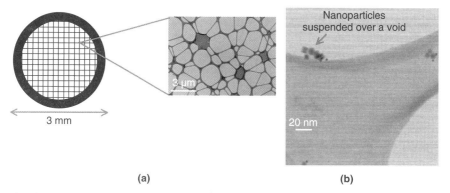

Fig. 4.45 Lacy carbon TEM sample grid. (a) Standard 3 mm TEM sample grid now used on all commercial instruments. The grid is coated with a support for the sample, which is transparent to electrons. This can be a continuous 100 nm thick carbon film or, as shown in the figure, a lacy carbon film particularly suited to imaging nanoparticles. (b) There is a high probability of some nanoparticles sticking to the edge of the lacy film so that they are suspended over a void as shown by the example image of Fe nanoparticles.

The electron beam is then directly incident on the particles with no intervening support.

An alternative method to obtain images using electrons is to focus the electron beam to a very small spot using magnetic lenses in a column similar to the focused ion beam column illustrated in Fig. 4.22 and raster this across the surface. A detector records the secondary electrons emitted at each position, and a map of the detected signal versus x and y is an image of the surface. This is the method used in a Scanning Electron Microscope (SEM). Contrast is obtained because different materials and topographical features emit secondary electrons at a different rate. The maximum resolution is defined by the size of the focused spot, which is ~5 nm. Figure 4.20 shows a high-quality SEM image of CoPt nanoparticles synthesized using EBL. TEMs can also be run in a rastered focused beam mode (scanning TEM or STEM) but commercial SEMs are simpler to operate, cheaper, and more flexible, being able to accept large samples with no restriction on the substrate material or thickness.

ADVANCED READING BOX 4.9—DIFFRACTION LIMIT OF MICROSCOPY

The ability of any optical system to resolve features is limited by diffraction of the illuminating particles (electrons, photons) around the aperture of the magnifying lens. An optically perfect lens with a finite diameter will not focus all particles to an infinitesimal point but to a diffraction pattern like the one shown below.

The central spot (the Airy disk) is the smallest feature that the microscope can produce in its image, and it has a radius given by

$$r = 1.22\lambda N,$$

where λ is the wavelength of the illuminating particles and N is the ratio of the focal length to the diameter of the lens producing the image (i.e., the f-number). The diffraction limit of the resolution of the microscope can be taken to be r. Generally $N \sim 1$, so the resolution of the instrument is in the range λ–2λ. Photons have a wavelength around 500 nm, so the resolution limit of an optical microscope is ~ 1 μm. The high quality of glass optics used in microscopes enables this limit to be reached.

The wavelength of electrons is given by the de Broglie formula, that is,

$$\lambda_e = \frac{h}{p},$$

where h is Planck's constant and p is the electron momentum. At typical electron energies used in a TEM, the electrons are traveling at more than half speed of light, so the relativistic mometum has to be used in the de Broglie formula. Thus,

$$\lambda_e = \frac{h}{\sqrt{2m_0 E \left(1 + \frac{E}{2m_0 c^2}\right)}},$$

where m_0 is the electron rest mass, E is the electron energy ($e \times$ accelerating voltage), and c is the speed of light. Accelerating potentials used in TEMs are typically in the range 50–500 keV, giving electron wavelengths in the range 0.005–0.001 nm, respectively. The aberrations in magnetic lenses are much greater than in glass optics, and the diffraction limit has never been observed in electron microscopy. In the best aberration-corrected instruments, resolutions ~ 0.08 nm have been achieved, which is good enough to resolve atoms.

PROBLEMS

1. A gas-phase nanoparticle source is producing Cu nanoparticles with a median diameter of 2 nm and depositing them on a substrate. The diameter of the deposited area is 2 cm. Given that Cu has a density of 8920 kg/m^3, determine the number of nanoparticles per second required from the source to produce a rate of increase of the deposited film thickness of 0.1 nm/sec.

2. A differential mobility analyzer (DMA) is used to filter particle sizes in an aerosol and has the dimensions shown in the figure. Determine the voltage that has to be applied to the central electrode for the device to filter 20-nm-diameter particles. Use the parameters in Advanced Reading Box 4.4 and assume a mean free path of the air molecules of 150 nm to evaluate the Cunningham correction.

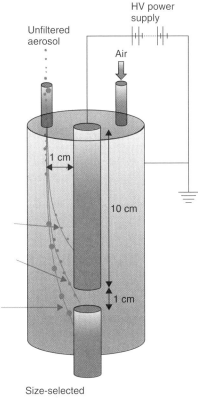

Size-selected
aerosol

3. In the chemical synthesis of FePt nanoparticles from $Pt(acac)_2$ and $Fe(CO)_5$, the process is started with 100 mg of $Pt(acac)_2$. How many mg of $Fe(CO)_5$ must be added to produce the equiatomic alloy stoichiometry of FePt?

4. A scanning tunneling microscope (STM) image is taken of a metal surface in constant current mode with a set current of 1 nA. Assuming that the bottom of the metal conduction band is 10 eV below the Fermi level and that the work function is 5 eV, determine the relative change in height of the tip above the surface if the set current is increased to 2 nA.

5. An AFM is used is to scan a surface in non-contact and constant height mode at a height, z, of 40 nm above the surface. The resonant frequency of the cantilever is 32 kHz and its stiffness constant is 1 N/m. Assuming a van der Waals attraction ($F(z) \propto 1/z^7$) between the tip and the surface, estimate the change in frequency produced by a feature on the surface with a height of 1 nm.

6. Compare the fundamental limit of resolution of imaging by X rays with a photon energy of 100 keV and electrons with an energy of 100 keV, assuming an f-number of 1.

REFERENCES

1. C. Binns, K. N. Trohidou, J. Bansmann, S. H. Baker, J. A. Blackman, J.-P. Bucher, D. Kechrakos, A. Kleibert, S. Louch, K.-H. Meiwes-Broer, G. M. Pastor A. Perez and Y. Xie, The behaviour of nanostructured magnetic materials produced by depositing gas-phase nanoparticles, *J. Phys. D: Appl. Phys.* **38** (2005), R1–R23.

2. K. O'Grady and A. Bradbury, Particle size analysis in ferrofluids, *J. Magn. Magne. Mater.* **39** (1983), 91–94.

3. K. Sattler, J. Mhlback, and E. Recknagel, Generation of metal clusters containing from 2 to 500 atoms, *Phys. Rev. Lett.* **45** (1980), 821–824.

4. S. H. Baker, S. C. Thornton, K. W. Edmonds, M. J. Maher, C. Norris, and C. Binns, Characterisation of a gas aggregation source for the preparation of size-selected sanoscale transition metal clusters, *Rev. Sci. Instrum.* **71** (2000), 3178–3183.

5. H. Haberland, M. Karrais, M. Mall, and Y. Thurner, Thin films from energetic cluster impact: A feasibility study, *J. Vacuum Sci. Technol. A* **10** (1992), 3266–3271.

6. T. G. Dietz, M. A. Duncan, D. E. Powers, and R. E. Smalley, Laser production of supersonic metal cluster beams, *J. Chem. Phys.* **74** (1981), 6511–6512.

7. P. Milani and W. A. deHeer, Improved pulsed laser vaporization source for production of intense beams of neutral and ionized clusters, *Rev. Sci. Instrum.* **61** (1990), 1835–1838.

8. W. Bouwen, P. Thoen, F. Vanhoutte, S. Bouckaert, F. Despa, H. Weidele, R. E. Silverans, and P. Lievens, Production of bimetallic clusters by a dual-target dual-laser vaporization source, *Rev. Sci. Instrum.* **71** (2000), 54–58.

9. H. R. Siekmann, C. Luder, J. Faehrmann, H. O. Lutz, and K. H. Meiwes-Broer, The pulsed arc cluster ion source (PACIS), *Z. Phys. D* **20** (1991), 417–420.

10. R. P. Methling, V. Senz,. D. Klinkenberg, T. Diederich, J. Tiggesbaumker, G. Holzhuter, J. Bansmann, and K. H. Meiwes-Broer, Magnetic studies on mass-selected iron particles, *Eur. Phys. J. D* **16** (2001), 173–176.

11. H. V. Tafreshi, P. Piseri, G. Benedek, and P. Milani, The role of gas dynamics in operation conditions of a pulsed microplasma cluster source for ?nanostructured thin films deposition, *J. Nanosci. Nanotechnol.* **6** (2006), 1140–1149.

12. S. H. Baker, S. C. Thornton, K. W. Edmonds, M. J. Maher, C. Norris, and C. Binns, The construction of a gas aggregation source for the preparation of size-selected nanoscale transition metal clusters, *Rev. Sci. Instrum.* **71** (2000), 3178–3183.

13. Bu. Wrenger and K. H. Meiwes-Broer, The application of a Wien filter to mass analysis of heavy clusters from a pulsed supersonic nozzle source, *Rev. Sci. Instrum.* **68** (1997), 2027–2030.

14. B. von Issendorf and R. E. Palmer, New high transmission infinite range mass selector for cluster and nanoparticle beams, *Rev. Sci. Instrum.* **70** (1999), 4497–4501.

15. C. Binns, Nanoclusters deposited on surfaces, *Surf. Sci. Rep.* **44** (2001), 1–49.

16. W. D. Knight, K. Clemenger, W. A. de Heer, W. A. Saunders, M. Y. Chou, and M. L. Cohen, Electronic shell structure and abundances of sodium clusters, *Phys. Rev. Lett.* **52** (1984), 2141–2143.

17. I. Katakuse, T. Ichihara, Y. Fujita, T. Matsuo, T. Sakurai, and H. Matsuda, Mass distributions of copper, silver and gold clusters and electronic shell structure, *Int. J. Mass Spectrom. Ion Processes* **67** (1985), 229–236.

18. P. Liu, P. J. Ziemann, D. B. Kittelson, and P. H. McMurry, Generating particle beams of controlled dimensions and divergence: I. Theory of particle motion in aerodynamic lenses and nozzle expansions, *Aerosol Sci. Technol*. **22** (1995), 293–313.

19. F. Di Fonzo, A. Gidwani, M. H. Fan, D. Neumann, D. I. Iordanoglou, J. V. R. Heberlein, P. H. McMurry, and S. L. Girshick, Focused nanoparticle-beam deposition of patterned microstructures, *Appl. Phys. Lett*. **77** (2000), 910–912.

20. K. Wegner, P. Piseri, H. Vahedi Tafreshi, P. Milani, Cluster beam deposition: a tool for nanoscale science and technology *J. Phys. D* **39** (2006), R439.

21. W. A. Saunders, P. C. Sercel, R. B. Lee, H. A. Atwater, K. J. Vahala, R, C. Flagan, and E. J. Escorcia-Aparcio, Synthesis. of luminescent silicon clusters by spark ablation, *Appl. Phys. Lett*. **63** (1993), 1549–1551.

22. S. L. Girshick, C.-P. Chiu, R. Muno, C. Y. Wu, L. Yang, S. K. Singh, and P. H. McMurry, Thermal plasma synthesis of ultrafine iron particles, *J. Aerosol Sci*. **24** (1993), 367–382.

23. S. E. Pratsinis, Flame aerosol synthesis of ceramic powders, *Prog. Energy Combust. Sci*. **24** (1998), 197–219.

24. T. Hyeon, Chemical synthesis of magnetic nanoparticles, *Chem. Commun*. (2003), 927–934.

25. S. Sun, C. B. Murray, D. Weller, L. Folks, and A. Moser, Monodisperse FePt nanoparticles and ferromagnetic nanocrystal superlattices, *Science* **287** (2000), 1989–1992.

26. E. M. Boatman, G. C. Lisensky, and K. J. Nordell, A safer, easier, faster synthesis for CdSe quantum dot nanocrystals, *J. Chem. Educ*. **82** (2005), 1697–1699.

27. see http://mrsec.wisc.edu/Edetc/nanolab/CdSe/

28. J. C. Garcia-Martinez, R. W. J. Scott, and R. M. Crooks, Extraction of monodisperse palladium nanoparticles from dendrimer templates, *J. Am. Chem. Soc*. **125** (2003), 11190–11191.

29. J. C. Garcia-Martinez and R. M. Crooks, Extraction of Au nanoparticles having narrow size distributions from within dendrimer templates, *J. Am. Chem. Soc*. **126** (2004), 16170–16176.

30. K. Esumi, Dendrimers for nanoparticle synthesis and dispersion stabilisation, in *Colloid Chemistry II*, M. Antomietti, ed., Topics in Current Chemistry, Vol. **227**, Chapter 2, Springer 2003.

31. S. Staniland, W. Williams, N. Telling, G. Van Der Laan, A. Harrison, and B. Ward, Controlled cobalt doping of magnetosomes in vivo, *Nature Nanotechnol*. **3** (2008), 158–162.

32. V. S. Coker, N. D. Telling, G. Van Der Laan, R. A. D. Pattrick, C. I. Pearce, E. Arenholz, F. Tuna, R. E. P. Winpenny, and J. R. Lloyd, Harnessing the extracellular bacterial production of nanoscale cobalt ferrite with exploitable magnetic properties, *Proc. Nat. Acad. Sci*. **3** (2009), 1922–1928.

33. R. E. Haufler, J. Conceicao, L. P. F. Chibante, Y. Chai, N. E. Byrne, S. Flanagan, M. M. Haley, S. C. O'Brien, B C. Pan, Z. Xiao, I W. E. Billups, M. A. Ciufolini, R. H. Hauge, J. L. Margrave, L. J. Wilson, R. F. Curl, and R. E. Smalley, Efficient production of C_{60} (buckminsterfullerene), $C_{60}H_{36}$ and the solvated buckide ion, *J. Chem. Phys*. **94** (1990), 8634–8636.

34. K. Kikuchi, N. Nakahara, T. Wakabayashi, M. Honda, H. Matsumiya, T. Moriwaki, S. Suzuki, H. Shiromaru, K. Saito, K. Yamauchi, I. Ikemoto, and Y. Achiba, Isolation and identification of fullerene family: $C_{76}, C_{78}, C_{82}, C_{84}, C_{90}$ and C_{96}, *Chem. Phys. Lett*. **188** (1992), 177–180.

35. For more details on the Nanobama synthesis, see http://www.nanobama.com/how/how.htm

36. M. C. Hersam, Progress towards monodisperse single-walled carbon nanotubes, *Nature Nanotechnol*. **3** (2008), 387–394.

37. X. Tu, S. Manohar, A. Jagota, and M. Zheng, DNA sequence motifs for structure-specific recognition and separation of carbon nanotubes, *Nature* **460** (2009), 250–253.

38. J. A. Rodríguez-Manzo, M. Terrones, H. Terrones, H. W. Kroto, L. Sun, and F. Banhart, In situ nucleation of carbon nanotubes by the injection of carbon atoms into metal nanoparticles, *Nature Nanotechnol*. **2** (2007), 307–311.

39. See, for example, http://www.swnano.com

40. I. Haller, M. Hatzakis, and R. Shrinivasan, High-resolution positive resists for electron beam exposure, *IBM J. Res. Dev*. **12** (1968), 251.

41. A, F. Grigorescu and C. W. Hagen, Resists for sub-20-nm electron beam lithography with a focus on HSQ: State of the art, *Nanotechnology* **20** (2009), 292001 (31 pp).

42. V. L. Mironov, B. A. Gribkov, S. N. Vdovichev, S. A. Gusev, A. A. Fraerman, O. L. Ermolaeva, A. B. Shubin, A. M. Alexeev, P. A. Zhdan and C. Binns, Magnetic force microscope tip induced remagnetization of CoPt nanodiscs with perpendicular anisotropy, *J. Appl. Phys*. **106** (2009), 053911 (8pp).

43. R. Clampitt, K. L. Aitken, and D. K. Jefferies, Intense field-emission ion source of liquid metals, *J. Vac. Sci. Technol*. **12** (1975), 1208–1208.

44. G. Taylor, Disintegration of water droplets in an electric field, *Proc. R. Soc. London Ser. A* **280** (1964), 383–397.

45. http://www.youtube.com/watch?v=zKX4G6tQOio

46. F. A. Stevie, D. P. Griffis, and P. E. Russell, *Introduction to Focused Ion Beams: Instrumentation, Theory, Techniques and Practice*, L. A. Gianuzzi and F. A. Stevie, eds., Springer, New York, 2005, Chapter 3.

47. A. Bachtold, M. Henny, C. Terrier, C. Strunk, C. Schönenberger, J.-P. Salvetat, J.-M. Bonard, and L. Forró, Contacting carbon nanotubes selectively with low-ohmic contacts for four-probe electric measurements, *Appl. Phys. Lett*. **73** (1998), 274.

48. A. Bachtold, C. Strunk, J.-P. Salvetat, J.-M. Bonard, L. Forró, T. Nussbaumer, and C. Schönenberger, Aharonov–Bohm oscillations in carbon nanotubes, *Nature* **397** (1999), 673–675.

49. J.-P. Cleuziou, W. Wernsdorfer, V. Bouchat, T. Ondarçuhu, and M. Monthioux, Carbon nanotube superconducting quantum interference device, *Nature Nanotechnol*. **1** (2006), 53–59.

50. J.-L. Li, J.-F. Jia, X.-J. Liang, X. Liu, J.-Z. Wang, Q.-K. Xue, Z.-Q. Li, J. S. Tse, Z. Zhang, and S. B. Zhang, Spontaneous assembly of perfectly ordered identical-size nanocluster arrays, *Phys. Rev. Lett*. **88** (2002), 066101.

51. A. Borg, A.-M. Hilmen, and E. Bergene, STM studies of clean, CO- and O_2-exposed Pt(100)-hex-R0.7°, *Surf. Sci*. **306** (1994), 10–20.

52. S.-C. Chin, Y.-C. Chang, and C.-S. Chang, The fabrication of carbon nanotube probes utilizing ultra-high vacuum transmission electron microscopy, *Nanotechnology* **20** (2009), 285307.

53. D. M. Eigler and E. K. Shweizer, Positioning single atoms with a scanning tunneling microscope, *Nature* **344** (1990), 524–526.

54. M. F. Crommie, C. P. Lutz, D. M. Eigler, and E. J. Heller, Waves on a metal surface and quantum corrals, *Surf. Rev. Lett.* **2** (1995), 127–137.

55. S.-W. Hla, Scanning tunneling microscopy single atom/molecule manipulation and its application to nanoscience and technology, *J. Vac. Sci. Technol. B* **23** (2005), 1351.

56. M. T. Cuberes, R. R. Schlittler, and J. K. Gimzewski, Room-temperature repositioning of individual C_{60} molecules at Cu steps: Operation of a molecular counting device, *Appl. Phys. Lett.* **69** (1996), 3016.

57. P. Moriarty, Y.-R. Ma, M. D. Upward, and P. H. Beton, Translation, rotation and removal of C_{60} on Si(100)-(2 × 1) using anisotropic molecular? manipulation, *Surf. Sci.* **407** (1998), 27.

58. S. Fujiki, K. Masunari, R. Nouchi, H. Sugiyama, Y. Kubozono, and A. Fujiwara, Nanoscale patterning by manipulation of single C_{60} molecules with a scanning tunneling microscope, *Chem. Phys. Lett.* **420** (2006), 82–85.

59. G. Binnig, C. F. Quate, and C. Gerber, Atomic force microscope, *Phys. Rev. Lett.* **56** (1986), 930–933.

60. F. J. Giessibl, AFM's path to atomic resolution, *Mater. Today* **May** (2005), 32–41.

61. F. J. Giessibl, S. Hembacher, H. Bielefeldt, and J. Mannhart, Subatomic features on the silicon (111)-(7 × 7) surface observed by atomic force microscopy, *Science* **289** (2000), 422–425.

62. R. D. Piner and C. A. Mirkin, Effect of water on lateral force microscopy in air, *Langmuir* **13** (1997), 6864–6868.

63. K. Salaita, Y. Wang, and C. A. Mirkin, Applications of dip-pen nanolithography, *Nature Nanotechnol.* **2** (2007), 145–155.

64. K. B. Lee, E. Y. Kim, C. A. Mirkin, and S. M. Wolinsky, The use of nanoarrays for highly sensitive and selective detection of human immunodeficiency virus type 1 in plasma, *Nano Lett.* **4** (2004), 1869–1872.

65. B. Fultz and J. M. Howe, *Transmission Electron Diffraction and Diffractometry of Materials*, Springer, New York, 2007.

CHAPTER 5

Single-Nanoparticles Devices

The focus so far has been on the special properties of nanostructures, or advanced materials formed from them, which is broadly covered by the term *incremental nanotechnology* defined in the introduction. In this chapter we will examine the possibility of making nano-devices built out of a single nanostructure. This can be, for example, a metal or semiconductor nanoparticle, a fullerene molecule, or a carbon nanotube "functionalized" to individually perform a task. The chapter thus delves into the world of evolutionary nanotechnology. Some examples of the kind of system we will be dealing with were briefly described in the introduction (see Fig. 0.3). The emphasis will be on electronic devices or data storage rather than on nano-mechanical systems, though there are some splendid examples of these based on individual nanostructures—for example, the nano-cheesewire described in Chapter 3 (see Fig. 3.20). The focus will be on three examples of functionalized particles—that is, magnetic nanoparticles for storing digital data, nanoparticles that act as electronic components for ultra-high density circuits, and light-emitting quantum dots. There are many additional examples, but the topics presented here will provide an understanding of what is meant by Evolutionary Nanotechnology. The nanoparticles discussed in this chapter can also be functionalized with biological molecules to perform a medical function, but the discussion of these is deferred to Chapter 6, which is entirely devoted to medical applications.

5.1 DATA STORAGE ON MAGNETIC NANOPARTICLES

No one can fail to be impressed by the massive increase in the density of stored data on magnetic disks achieved over the last few decades. On magnetic hard disks, digital data are stored as a series of binary digits or bits ("1" or "0") patterned by reversals of magnetization in circular tracks around the surface of a rotating disk and the density is normally quoted in bits/in.^{2}. This is the one aspect of nantechnology that is stuck with imperial units, probably related to the fact that the hard disk industry is dominated by the United States. The first hard disk magnetic storage system the RAMAC produced by IBM in 1956 had a storage density of 2000 bits/in.^{2}. Fifty years later, personal computer hard disks had typical storage densities of 200 Gb/in.^{2} (2×10^{11} bits/in.^{2}), an increase by a factor of 100,000,000. Recently, Seagate demonstrated a disk system with a storage density of 421 Gb/in.^{2}, and existing magnetic technology is predicted to "top out" at about 1 Tb/in.^{2} (10^{12} bits/in.^{2}). On the RAMAC device a data bit required a storage area of side $\sim$0.5 mm, which has shrunk to about 35 nm. To put the increase in perspective, if it took a football field to store a single data bit in 1956, it now takes an area the size of a lentil. This push to ever-higher storage densities is driven by demand because we are increasingly used to storing and transferring massive amounts of data as multimedia uses ever-higher definition pictures and movies. Until now the technology has kept pace with this demand, but the storage density is starting to come up against fundamental limits that require a paradigm shift in storage methodology.

In the last few years there have appeared many different types of mass storage devices including flash drives, rewritable DVDs, and so on, whose storage density rivals magnetic recording. These will not be discussed in this section, which will focus on the fundamental limits of magnetic storage and describe how nanotechnology can overcome these. Similar limits are to be found in the other storage technologies with, again, nanotechnology offering solutions. In Section 5.3 we will examine the prospects for miniaturizing electronics using nanoparticles and carbon nanotubes as circuit elements. Magnetic recording of data has some distinct advantages—for example, its extreme nonvolatility. The magnetization of a piece of magnetic material can be reversed at gigahertz frequency for geological timescales without affecting it in any way. Solid-state devices such as flash memories have a limited number of read/write cycles before storage becomes unreliable. The main disadvantage of hard disk drives at present is their complex mechanical structure that makes them less robust than flash drives, but in future generations of magnetic recording devices it may be possible to dispense with the rotating disk.

At present, data on hard disks are stored on a continuous magnetic film consisting of densely packed nanoparticles on the surface with data bits written onto small areas containing many particles (see Fig. 5.1). Suppose, instead, that data could be stored by magnetizing *individual* magnetic nanoparticles such as the Fe particles with a diameter of 3 nm shown in Fig. 1.9. If these were in an array and separated sufficiently so that changing the magnetization of one did not affect its neighbors (separation required is approximately equal to particle diameter), then, for example, "up" magnetization could represent "1" and "down" magnetization could represent "0" (see Fig. 5.2). The recording density in such as system could reach 100 Tb/in.^{2}—that is, a factor of 100 higher than the predicted limit of current technology. In our hypothetical single-particle recording system, each nanoparticle becomes a memory element and performs an individual function—that is, storing a single data bit—and the technology is thus evolutionary nanotechnology. This task requires a pair of transistors in a flash memory, for example. We will discuss below the fundamental limits of magnetic recording and what problems have to be overcome to enable data storage onto individual nanoparticles.

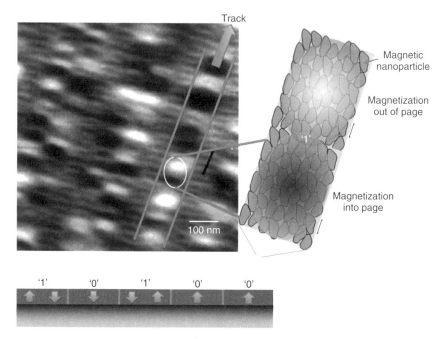

Fig. 5.1   MFM Image of 394 Gb/in² disk. Magnetic Force Microscope (MFM – see Chapter 4, Section 4.4.4) image of the magnetization pattern on the surface of a 500 Gigabyte Seagate 5400.6 Momentus hard disk. The magnetic medium is granular and consists of nanoparticles with sizes in the range 10–20 nm. The information is stored by magnetizing the medium perpendicular to the disk surface (in and out of the page). As illustrated in the lower diagram A magnetic reversal within the area of a data bit represents a '1' and the absence of a reversal represents a '0'. Each data bit is written onto a number (∼100) of the nanoparticles as indicated by the blow-up showing an individual '1' bit. MFM image reproduced with permission from NanoMagnetics Instruments (www.nanomagnetics-inst.com).

In order to permanently store data, a piece of magnetic material has to be a single magnetic domain, which, as discussed in Chapter 1, Section 1.2, requires it to have a diameter less than about 100 nm. Thus any particle we would be interested in down to the size of a single atom can be assumed to be a single magnetic domain. As we shrink the size down from the single-domain limit at about 100 nm toward single atoms, there is a size below which the magnetization becomes unstable due to thermal fluctuations, and the magnetization will be lost in the absence of an applied magnetic field. The critical size, below which the particle demagnetizes itself, depends on a number of factors including the temperature, shape, and chemical composition of the particle; but at room temperature for spherical Fe nanoparticles, for example, it is of the order of 6 nm (see Advanced Reading Box 5.1). The temperature below which particles of a given size have a stable magnetization direction and can thus store data is known as the "blocking temperature."

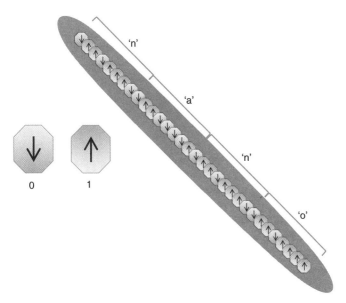

Fig. 5.2 Hypothetical data storage onto individual magnetic nanoparticles. Schematic of a recording medium where each nanoparticle stores a data bit with, for example a '0' represented by 'down' magnetization and a '1' by 'up' magnetization. The image shows the word 'nano' stored in ASCII code along a line of particles. If these particles were at the superparamagnetic limit (see Advanced Reading Box 5.1) the storage density would be ~ 100 Tb/in^2, that is, about 100 times higher than the predicted limit for existing technology.

ADVANCED READING BOX 5.1—SUPERPARAMAGNETIC LIMIT

In general, magnetic nanoparticles can be assumed to have a uniaxial anisotropy; that is, the magnetic energy is lowest when the particle magnetization points in either direction along a single "anisotropy axis" as indicated on the left of the figure.

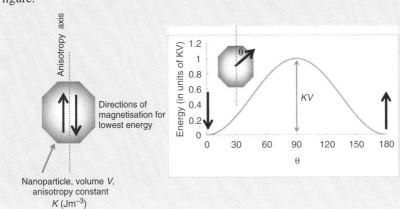

The strength of the anisotropy is given by the anisotropy constant K with units J/m^3. The difference between maximum and minimum magnetic energies (in this case the difference in energy between the magnetization pointing along the anisotropy axis and perpendicular to it) is given by KV, where V is the volume of the particle. Thus plotted as a function of the angle between the direction of magnetization and the anisotropy axis, θ, the extra energy arising from the anisotropy, ΔE, is zero when the magnetization points along the anisotropy axis in either direction and is maximum, at a value KV, when it is perpendicular. It is found in practice that

$$\Delta E = KV \sin^2 \theta.$$

Thus the particle magnetization rests in one of two perpendicular directions given by $\theta = 0$ and 180° (as depicted in Fig. 5.2), and there is an energy barrier of height KV to surmount in order to reverse the magnetization. This is the anisotropy barrier. As the particle gets smaller, so does KV; and for a sufficiently small particle the anisotropy barrier will be comparable to the thermal energy kT. This means that thermal fluctuations are sufficient to drive the particle magnetization from one state to the other, and the particle will not stay magnetized. The temperature below which the particle magnetization is stuck (on some defined timescale) in one direction or the other is known as the blocking temperature and is calculated below.

The mean time, τ, that a particle spends in one of the two minimum energy directions (the 'lifetime') is given by the Arrhenius law:

$$\tau = \tau_0 \exp\left(\frac{KV}{kT}\right),$$

where τ_0 is the natural lifetime, that is, the lifetime at the high temperature limit. This has been measured to be about 1 nsec for nanoparticles, but its magnitude does not significantly affect the blocking temperature. The lifetime is only infinite at $T = 0$, so to specify a blocking temperature, we must decide on what timescale the moment is stuck. For example, we could use the recording industry norm that the magnetization is stable for 40 years $= 1.26 \times 10^9$ sec; then the blocking temperature is

$$T_B = \frac{KV}{k(\log \tau - \log \tau_0)} = \frac{KV}{k(\log(1.26 \times 10^9) - \log(10^{-9}))}$$

So putting in some numbers, let's say we have a 2-nm-diameter Fe nanoparticle with an anisotropy constant of 2×10^5 J/m^3 (a value measured for 2-nm Fe particles [1]). The blocking temperature is 15 K. If we wanted to reliably store data on Fe nanoparticles at room temperature, they would have to have a diameter of at least 6 nm. In order to store data on yet smaller particles we need to find ways of increasing K. This can be done by using high anisotropy alloys such as FePt or core-shell particles (see text).

This is a fundamental limit that sets the maximum data storage density at a given temperature. Particles below the critical size are "superparamagnetic"; that is, they will magnetize in an applied field along the direction of the field but will demagnetize when the field is turned off. In fact, this is normal paramagnetic behavior and the "super" pronoun is used because the magnetic moment of the nanoparticle is derived from the individual atomic magnetic moments locked together to form a single "giant" moment. The limit to the data storage density set by the onset of thermal instability in the nanoparticle magnetic moment is known as the superparamagnetic limit. This would set the ultimate limit for magnetic storage at about 100 Tb/in.^{2} unless the temperature is reduced, but cooling the storage medium is generally disregarded. This is not because it is technologically difficult but because it is pointless to store large quantities of data in a tiny space if a large cooling unit is required next to the storage device. The 1 Tb/in.^{2} quoted above for current technology is also a superparamagnetic limit, but present-day devices store a single data bit on an area of a disk containing about 100 nanoparticles (see Fig. 5.1); so if they are individually superparamagnetic, information can't be stored.

Achieving magnetic storage at data densities significantly higher than 1 Tb/in.^{2} requires technology that can read from and write to individual nanoparticles. This involves two major and presently unsolved technological challenges. One is the ability to organize suitable nanoparticles (i.e., those that can hold a permanent magnetization at room temperature) in an ordered array such as lines of uniformly spaced nanoparticles as illustrated in Fig. 5.2. The other is a method of reading the magnetization of individual nanoparticles and changing the magnetization of a selected particle (the "write" process).

A promising approach to producing suitable ordered arrays is to chemically prepare nanoparticles of FePt, a magnetic alloy which, in the correct phase, has a very high anisotropy and thus, for a given size, a high blocking temperature (see Advanced Reading Box 5.1). The basic recipe for the synthesis of FePt nanoparticles in a liquid suspension was described in Chapter 4, Section 4.1.7 (see Figs. 4.11 and 4.12). Ordered arrays of the FePt nanoparticles can be produced by dispensing a drop of the suspension on a flat surface and allowing the liquid to evaporate. With this method the spacing between the particles and ordered arrangement can be controlled by choosing the surfactant coating of the particles as described in Section 4.1.7. Figure 5.3 shows an atomic force microscope (AFM) image of 6-nm FePt nanoparticles manufactured by chemical synthesis and formed as an ordered square array. These are an illustrative example of functionalized nanoparticles. The constituent alloy of the particles has been chosen to give them suitable magnetic properties, and their surface has been coated with

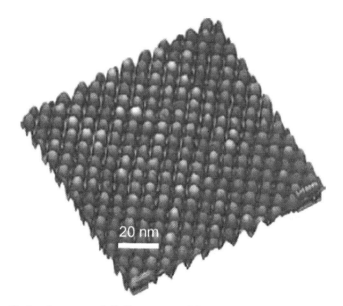

Fig. 5.3 Ordered array of FePt nanoparticles. AFM image of ordered array of 6 nm diameter FePt nanoparticles synthesized using the method described in section 4.1.7. Reproduced from the nanowerk website (http://www.nanowerk.com/spotlight/spotid=301.php) with permission from Prof. Shouheng Sun, Brown University, U.S.A.

a surfactant that enables them to self-assemble with the desired arrangement and spacing. It is also possible to control the magnetic anisotropy of the nanoparticle by coating each one with a thin shell a few atomic layers thick of an antiferromagnetic material such as Cr. This can be done using the gas-phase production method described in Chapter 4, Section 4.1.1 and adds a further layer of functionalization allowing the synthesis of complex nanoparticles with "designed" properties and function.

The problem of producing ordered arrays of nanoparticles can therefore be solved in principle, though doing it in a reliable way with the right sort of nanoparticles over macroscopic areas remains a challenge for bottom-up methods. It is also possible to manufacture magnetic arrays using top-down methods such as electron beam lithography (EBL; see Fig. 5.4) and focused ion beam (FIB) milling (see Chapter 4, Section 4.2.2). The problem with these methods is that they have not yet achieved the synthesis of sub-10-nm nanostructures possible with bottom-up methods. In addition, they are exceedingly slow at manufacturing macroscopic areas of nanoparticles and would be unsuitable for the commercial manufacture of storage media. They can be used, however, to make test areas of nanoparticles to demonstrate the principle of single-particle magnetic storage as discussed below.

Even if producing ordered magnetic nanoparticle arrays over macroscopic areas were routinely achievable, reading from and writing to the array is at present not feasible on anything like the scale shown in Fig. 5.3. The process of

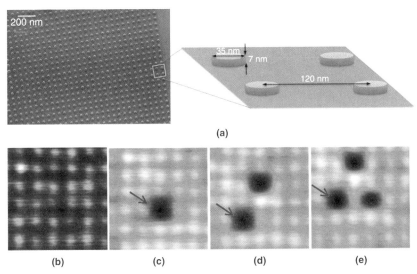

Fig. 5.4 **Writing to CoPt nanoparticle array using MFM.** Array of CoPt nanodisks 35 nm in diameter, 7 nm thick with a spacing of 120 nm synthesized on a Si substrate using EBL (see Chapter 4, Section 4.2.1). (b) 700 nm × 700 nm MFM scan of array with every nanodisk initially magnetized in the same direction (opposite to magnetization of the MFM tip. (c) − (e) Successive reversal of the magnetization of 3 chosen particles by lowering the tip towards them and then withdrawing to perform the MFM scan. The first particle reversed is indicated by a red arrow in each frame. The space occupied by each bit corresponds to a storage density of $\sim$40 Gb/in^2. Reproduced with the permission of the American Institute of Physics from V. L. Mironov et al. [4].

reading and writing (R/W) requires the capability of imaging the magnetic pattern on the surface with the resolution of a single nanoparticle and changing the magnetization state of a single chosen nanoparticle. At present the method for imaging magnetic patterns with the highest resolution is magnetic force microscopy (MFM; see Chapter 4, Section 4.4.4); this has demonstrated a resolution of about 20 nm in ambient conditions [3], but this is certainly not routine. It is likely that better resolution can be obtained in vacuum; but as with cooling, this is not an option for any commercial storage device.

The MFM would also be the best instrument to achieve writing to the array by using the ultra-sharp magnetic tip lowered toward a single nanoparticle to flip its magnetization. A careful study of this write process was carried out by V. L. Mironov et al. [4] using an array of CoPt nanoparticles manufactured by EBL. This material has a perpendicular magnetic anisotropy; thus the particles are magnetized either "up" or "down," making them suitable to store digital data. The sample is shown in Fig. 5.4a and consists of a square array of nanodisks 35 nm in diameter, 7 nm thick, and spaced 120 nm apart on a silicon substrate. Initially the entire array was magnetized in the same direction and opposite to the tip magnetization by an applied magnetic field. An MFM scan with the cantilever at a height of ~50 nm above the surface in this initial state (Fig. 5.4b) shows a uniform array of white dots. The magnetization of selected particles was then reversed by moving the MFM tip over them and lowering it down to touching. Moving the tip back to scanning height and taking an MFM image then showed the new reversed state of the particle. The sequence Fig. 5.4c–Fig. 5.4e shows the successive remagnetization of three particles, that is, 3 bits "written."

There is one nanodisk, that is, 1 bit per 120-nm × 120-nm area; this corresponds to a data storage density of about 40 Gb/in.^{2}, so this technology has yet to achieve the density currently available in commercial devices. The important point is that there is no fundamental limit why this process should not be able to achieve much higher densities and "leap-frog" past the fundamental limits that will limit conventional magnetic storage technology—at least in terms of density.

The remaining problem is the read/write speed, or data transfer rate (DTR), which, for single MFM tips, is extremely slow. IBM has put a great deal of investment into an atomic force microscope (AFM; see Chapter 4, Section 4.4.4)-based data storage system in which the information is stored by using the AFM tip to print tiny nanoscale dents into a polymer film. The data transfer rates for this method using a single tip can be 100,000 bits/sec[5], but this is still nearly 10,000 times slower than the DTR on a commercial hard disk drive. A possible solution is to use arrays of multiple AFM tips that work in parallel; in 1996, the first tentative steps in this direction were taken by Minne et al. [6], who used a double cantilever to produce nanoscale features on a Si surface by independently using the tips to produce a local spot of oxide. A significant leap in sophistication was reported three years later by M. Lutwyche et al. [7], who used micro-engineering methods to manufacture a 5 × 5 array of cantilevers shown in Fig. 5.5. The optical system used on a normal single-cantilever AFM (Chapter 4,

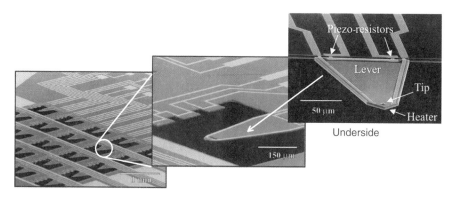

Fig. 5.5 Array of 5 × 5 AFM cantilevers for parallel operation. Multiple AFM cantilevers in a 5 × 5 array for parallel operating in read/write applications. Each cantilever has a built-in detection system to measure cantilever deflections (the pair of piezo resistors at the base) and some have a built-in heater to bend the cantilever towards the surface for writing. Reproduced with the permission of Elsevier Science from M. Lutwyche et al. [6].

Fig. 4.37) would not work on multiple-cantilever systems, so each cantilever in the array has a built-in detection system. This is the pair of weak link "piezo-resistors" at the base of each cantilever, which detect tiny cantilever deflections by changing resistance in response to changing strain. Each cantilever can also have a built-in heater (shown in the final image in Fig. 5.5) to bend it toward the surface to achieve writing though in this first device not every cantilever was fitted with one. It is an impressive piece of micro-engineering, but the field has moved on rapidly in the last decade and the state of the art is now 1024 cantilevers working in parallel on the IBM "millipede" project [8] with the next generation planned to include 4096 tips.

The same multiple-tip technology could be adapted for parallel operation MFM writing onto magnetic nanoparticles as demonstrated in Fig. 5.4. One point to bear in mind is that the AFM or MFM tips are a long way apart compared to the separation of the nanoparticles in the medium; thus consecutive bits in the data stream would not be stored on neighboring particles, but the software control system on any driver of such a storage system could easily handle this complication.

In summary, the state of the art in data storage on individual magnetic nanoparticles is at various stages relative to conventional magnetic storage technology, depending on which parameter is compared. Media of arrays of individual magnetic nanoparticles such as the one shown in Fig. 5.3 can be prepared with a density of >10 Tb/in.^{2}—that is, about a factor of $50\times$ greater than existing hard disk drives but not reliably over macroscopic areas. At present there is no way to read and write using a magnetic array at this density, and the state of the art in R/W to single nanoparticles is shown in Fig. 5.4. The storage density in this demonstration was about 40 Gb/in.^{2}—that is, an order of magnitude less than conventional technology and with a DTR several orders of magnitude lower. The key point, however, is that the limitations are technical rather than fundamental and can be overcome with improvements in MFM technology and multiple-tip arrays. There is no reason, in principle, why it should not eventually be possible to use arrays as dense as the one shown in Fig. 5.3. One further benefit of MFM-based technology is that it dispenses with the rotating disk and can be built entirely into a solid-state and extremely thin platform.

5.2 QUANTUM DOTS

In general, quantum dots are any particles that are sufficiently small that the electron energy level spacing becomes larger than thermal energy—*the quantum size effect*. The name is usually reserved, however, to describe nanoparticles of semiconductor materials that, when stimulated, emit light much like a nanoscale light-emitting diode (LED). One of their characteristics is that the wavelength of the light they emit depends on their size and shape, allowing control of the color of their fluorescence over a useful range. In liquid suspensions they have a number of important applications in biomedical research and diagnostic medicine, and these are described in detail in Chapter 6, Section 6.3.3. Before examining the light-emitting properties of semiconductor nanoparticles, it is worth briefly reviewing how bulk semiconductors fluoresce.

The electronic and optical behavior of a semiconductor is primarily controlled by an electron energy gap between the filled valence band and the empty conduction band. Unless electrons are somehow stimulated and given enough energy to cross the gap, the material is an insulator. The gap is sufficiently narrow that at room temperature there is enough thermal energy to excite a few electrons into the conduction band and the material attains a very low conductivity. The gap is also the reason they fluoresce because an electron excited across the gap by, for example, absorbing a sufficiently energetic photon will leave a hole in the valence band but will recombine with it a short time later, releasing the gap energy as a photon. The process is illustrated in the three steps shown in Fig. 5.6. Initially, an incoming photon promotes an electron across the gap, leaving a positively charged hole in the valence band, but the electron and hole interact with each other due to their opposite charges and form a bound pair known as an exciton (Fig. 5.6a). The absorbed photon can have any energy greater than the gap energy, and the electron can be excited well up into the conduction band. Although the electron can recombine with the hole from where it find itself, this is highly unlikely and during the lifetime

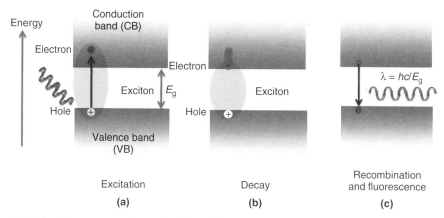

Fig. 5.6 **Fluorescence from a bulk semiconductor.** (a) An electron is promoted from the valence band to the conduction band by absorbing a sufficiently energetic photon. The electron interacts with the positively charged hole left behind to form a bound particle pair known as an exciton. (b) The electron energy decays down through the continuum of energy levels in the conduction band till it reaches the gap edge. (c) The electron recombines with the hole and the energy is released as a fluorescent photon with the gap energy. Thus a wide range of photon energies can be absorbed but mostly a single wavelength is emitted. In fact in stage (b) the electron will also decay through the excitonic energy levels below the bottom of the conduction band (see Advanced Reading Box 5.2) but in most cases these are very close to the bottom of the conduction band and the energy of the fluorescent photon is not significantly affected.

of the exciton the electron energy decays through the continuum of conduction band states until it is at the bottom at the gap edge (Fig. 5.6b). From here the exciton collapses; that is, the electron and hole recombine and the energy released (the gap energy) is given off as a photon of light (Fig. 5.6c). Thus semiconductor fluorescence is characterized by the absorption of a wide range of photon energies but emission at a single wavelength given by the gap energy.

The same basic process happens in nanoscale semiconductor particles or quantum dots; but in order to understand the size-dependence of the optical properties, we need to look in a little more detail at the energy of the recombination illustrated in Fig. 5.6c. While it is in existence the exciton, which is a bound positive and negative charge, behaves like a miniature atom with its own set of energy levels superimposed onto those of the background semiconductor (see Advanced Reading Box 5.2). So when the electron decays down through the conduction band states and reaches the gap edge, it then find itself at the top of another staircase of energy levels belonging to the exciton. It decays to the bottom of those before recombining as illustrated in the figure in Advanced Reading Box 5.2. In the bulk the exciton energies are very small, and the energy released when the exciton collapses is close to the bulk bandgap energy.

ADVANCED READING BOX 5.2—EXCITON STATES IN BULK SEMI-CONDUCTORS AND QUANTUM DOTS

When an exciton is formed in a bulk semiconductor as in Fig. 5.6a, the electron and the positively charged hole form a bound particle pair whose Hamiltonian is the same as for a hydrogen atom, apart from the masses of the constituent particles. The total mass of the exciton is

$$M = m_e^* + m_h^* \quad \text{and its reduced mass is} \quad \mu = \frac{m_e^* m_h^*}{m_e^* + m_h^*},$$

where m_e^* and m_h^* are the effective masses of the electron and hole. Since the problem is the same as for the hydrogen atom, we do not need to repeat the quantum mechanical calculation and we can use the standard result for the energy levels of the hydrogen atom modified by the change in mass of the bound pair and also take into account that the particles are interacting not through vacuum but through a medium with a dielectric constant ε_r relative to vacuum. The energy level series equivalent to the Rydberg series of the hydrogen atom is

$$E_n = E(\infty) - \frac{R_y^*}{n^2},$$

where $E(\infty)$ is the continuum state, which in this case is the energy of the bandgap, E_g, n is the principal quantum number ($n = 1, 2, 3 \ldots \ldots$), and R_y^* is the constant:

$$R_y^* = \left(\frac{\mu}{m_0 \varepsilon_r^2} \right) R_y,$$

where Ry is the Rydberg energy (13.6 eV). The exciton Rydberg series is illustrated in the figure. It is an extra series of energy levels lying just below the bottom of the conduction band.

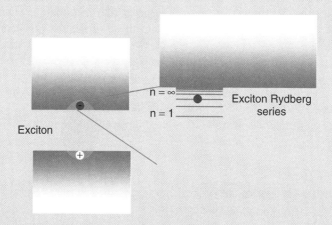

The exciton has a characteristic size (exciton Bohr radius) that can be related to the Bohr radius of the hydrogen atom, r_H (=0.053 nm), by

$$r_{ex} = r_H \varepsilon_r \frac{m_0}{\mu},$$

where m_0 is the rest mass of the electron. For example, the semiconductor CdSe has $\varepsilon_r = 10.1$, $m_e^* = 0.13 m_0$, and $m_h^* = 0.45 m_0$, giving an exciton Bohr radius, r_{ex}, of 5.3 nm. If the physical extent of the semiconductor is of the order of or less than r_{ex}, the exciton is "squeezed" and the excitonic energy levels are modified. Below a critical size the exciton enters a new regime where the energy is almost entirely determined by the confinement of the exciton within a box the size of the quantum dot. The energy between the first two energy levels of this system is given by

$$E = E_g + \frac{\hbar^2 \pi^2}{2 \mu r^2},$$

where, as before, E_g is the gap energy in the bulk semiconductor. The primary gap energy and thus the energy of the fluorescent photons is increased by a

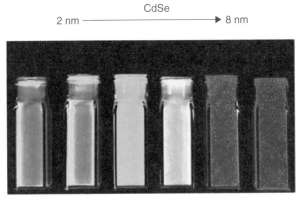

Fig. 5.7   Fluorescence from CdSe quantum dots of different sizes. Change in wavelength of fluorescence as a function of size from CdSe quantum dots excited with UV light. Note how the wavelength is reduced (bluer emission) as the size of the quantum dot is reduced. Reproduced with permission from Prof. Bawendi, Department of Chemistry, MIT.

factor proportional to $1/r^2$ from the bulk baseline as the size of the quantum dot is reduced. This is clearly seen in Fig. 5.7, showing the change in the color of the fluorescence of CdSe quantum dots excited by UV light as a function of their size.

There is a size associated with the exciton known as the exciton Bohr radius, which in the bulk is typically 10 nm. Thus when the spatial extent of the semiconductor is less than this, the exciton is "squeezed" and its energy levels change. It is no longer appropriate to think of it as a positive and negative charge bound within a larger system but as a particle in a box, whose size is that of the quantum dot. This is a classic problem in quantum mechanics; and it is well known that as the box is shrunk, the energy levels of the entrapped particle move apart. In this situation, the first exciton state, from which the electron recombines with the hole to produce fluorescence, is at a greater energy than the bandgap energy of the bulk. As the dot size is decreased, the exciton energy increases and the wavelength of the fluorescent photon is shorter ("bluer"). In a simple analysis (see Advanced Reading Box 5.2) the increase in energy of the fluorescent photon above the bulk semiconductor gap energy is proportional to $1/r^2$, where r is the radius of the dot. This provides a simple synthesis parameter—that is, the size of the dot—to control the colour of the fluorescence. This is nicely demonstrated in Fig. 5.7, which shows the fluorescence from CdSe quantum dots of various sizes manufactured by the group of M. Bawendi at the Department of Chemistry, MIT. The synthesis of these dots is described in detail in references [26] and [27] in Chapter 4.

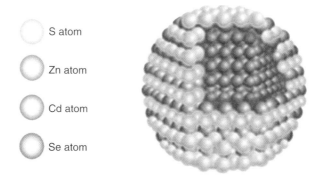

S atom

Zn atom

Cd atom

Se atom

Fig. 5.8   CdSe/ZnS core-shell quantum dot. Coating a CdSe quantum dot with ZnS prevents the surface reconstruction of the CdSe surface that leads to the creation of additional electronic states in the bandgap. These promote the relaxation of the excited electrons via non-radiative transitions and decrease the quantum yield of the dot.

Most quantum dots manufactured in liquid suspensions are compound semi-conductors such as CdSe due to their ease of manufacture. The naked CdSe nanoparticle, however, produces relatively weak and unstable emission. This is a common feature of all quantum dots and is due to the fact that the terminated crystal structure at the surface of the dot is unstable against a restructuring of the surface atoms into a new atomic arrangement known as a surface reconstruction (see Chapter 4, Section 4.4.1). The new structure introduces new electronic states in the bandgap (known as traps) that promote the relaxation of the excited state represented in Fig. 5.6b via nonradiative transitions rather than relaxation across the gap to produce a photon. A solution is to coat the CdSe with a shell of ZnS (Fig. 5.8), which has a similar crystal structure but a wider bandgap. The shell is not fluorescent, but it enables the CdSe core to maintain its crystal structure across the interface without restructuring so the number of emitted photons per absorbed photon, that is, the *quantum yield* of the dot is increased.

A method of manufacturing semiconducting quantum dots on surfaces that has been developed since about 1990 is to vacuum deposit a vapor of the compound semiconductor InAs onto GaAs substrates. The atomic lattice mismatch between the two compunds causes the InAs to self-assemble into nanoscale islands that show quantum dot behavior. A further example is short sections of semiconducting carbon nanotubes. The properties of nanotube quantum dots will be presented below.

5.3   NANOPARTICLES AS TRANSISTORS

One of the technological leaps in evolutionary nanotechnology has been the development of single nanoparticles that can function as transistors, which are the electronic building block of all digital electronic devices. A transistor is a three-terminal device whose conductivity can be controlled by an external voltage, and

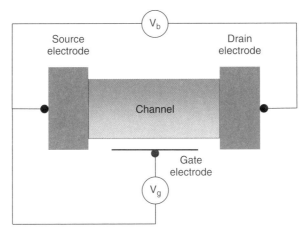

Fig. 5.9   Field-Effect Transistor (FET). Basic configuration of an FET. With no gate voltage applied the channel has a very high resistance so the bias voltage, V_b, between the source and the drain electrodes does not produce current. If a voltage V_g, relative to the source and drain potential, is applied to the gate electrode, above a threshold value the channel becomes conducting.

a basic schematic is shown in Fig. 5.9. Specifically, this is the configuration for a field-effect transistor (FET), but this is now the standard component in all modern digital and analog electronics. The voltage applied across the device—that is, between the source and drain electrodes—is normally termed the bias voltage, V_b. The gate is a terminal without a direct conducting electrical contact to the channel and influences the flow of electrons through the channel by the electric field generated when a voltage is applied to it. With no applied gate voltage, V_g, the channel has a very high resistance, but above a threshold gate voltage the channel conductance increases by many orders of magnitude. In essence the device is a voltage-controlled switch, and this simple building block is at the heart of the most complex electronic devices.

Like magnetic recording, the microelectronics industry has achieved amazing feats of miniaturization and modern computer chips contain more than 500 million FETs, which compares with 3500 on the Intel 8008 microprocessor manufactured in 1972. Figure 5.10 shows how the number of transistors in microprocessor chips has increased with years of manufacture on a logarithmic scale and demonstrates the exponential growth. It was Gordon Moore, a co-founder of Intel Corporation, who originally noticed the doubling in the number of transistors in integrated circuits every 2 or 3 years, and the exponential growth has since been dubbed *Moore's Law*. A significant driver for this remorseless drive to higher densities is commercial because in the industry the cost of processing a silicon wafer tends to stay the same; thus the more chips that can be included, the lower the cost per unit.

The increase in the number of transistors is achieved by reductions in the feature size patterned onto the chips. The width of the channel on an FET in

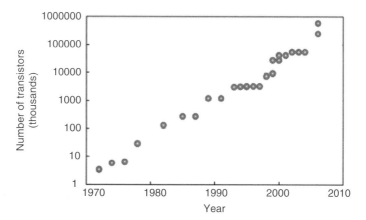

Fig. 5.10   Moore's Law. Growth in the number of transistors per microprocessor chip for devices manufactured by Intel Corporation since 1972. The exponential growth (doubling of components every 2 or 3 years) in integrated circuits was first noticed by Gordon Moore, co-founder of Intel) in 1965 and has been dubbed *Moore's Law*.

a modern chip is less than 65 nm, and the number of electrons applied to the gate is just thousands. The roadmap of the microelectronics industry refers to devices with gate lengths of 13 nm by 2013, below which the quantum size effect described in Section 5.2 will start to affect circuit behavior. Clearly, electronics is heading toward a transistor that is composed of a single nanoparticle in which the laws of quantum mechanics are exploited to process electrons one at a time; this transistor is referred to as a single electron transistor (SET). It is already possible to build such a device exploiting an effect known as *Coulomb blockade*.

The effect is illustrated in Fig. 5.11, which shows the energy states of a nanoparticle connected to source and drain electrodes via thin insulating barriers with a gate electrode in close proximity. The diagram has the rather odd but revealing structure of energy diagrams such as this in that energy is plotted vertically and distance horizontally. How such a device can be constructed in practice is discussed below, but for the moment let's assume it has been built. The nanoparticle contains N electrons and, as shown in the figure, the electron energy levels are discrete because the quantum size effect, introduced in Section 5.2, has become important. There are two electrons, one for each of the two allowed electron spins, in each quantum state. The top filled state will approximately align with the Fermi level—that is, the highest filled state of the continuum of states—of the electrodes. For the N electrons in the nanoparticle there are empty states (shown in gray) continuing up in energy from the highest filled state; however, these are not available to an electron from outside the particle. If we were somehow to force the nanoparticle to accept an extra electron to give it a population of $N + 1$, its negative charge would increase the energy of the particle by an amount $e^2/2C$, where C is its capacitance (see Advanced Reading Box 5.3). For any macroscopic object this term would be insignificant,

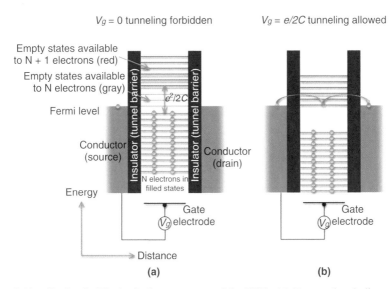

Fig. 5.11   Coulomb Blockade in a nanoparticle FET. (a) Energy level diagram of nanoparticle with discrete states due to the quantum size effect connected to source and drain electrodes via thin insulating tunnel barriers and a gate electrode in close proximity. The highest filled state of the particle approximately aligns with the highest filled electron level of the continuum in the source and drain electrodes (the Fermi level). The gray empty states are available to the N electron population of the nanoparticle but not to $N + 1$ electrons. If an extra electron was forced onto the particle the extra energy due to its charge ($e^2/2C$, where C is the capacitance – see Advanced Reading Box 5.3) would shift the available empty states to the red ones so tunneling from the source onto the particle is forbidden. If a voltage e/2C is applied to the gate electrode the lowest red state is pulled down into line with the Fermi level and if a bias is applied between the source and drain a single electron can tunnel onto the particle and off again. The device thus acts as an FET with the gate controlling whether tunneling conductance is allowed (one electron at a time), or not. To conduct two electrons at a time would require a further increase in the gate voltage.

but the capacitance of the particle is so small that the energy is significant compared to the normal energy level spacing. Thus for a particle containing $N + 1$ electrons the empty energy levels available are actually the ones shown in red in Fig. 5.11, which are shifted upwards by the capacitance energy. This means that the process of quantum mechanical tunneling is forbidden to electrons from the source electrode. Applying a voltage of $e/2C$ to the gate electrode (Fig. 5.11b) brings the lowest red state into line with the source electrode, and applying a bias voltage between the source and drain will enable a single electron to hop onto the nanoparticle and then off. At this gate voltage, there would only be one electron at a time allowed onto the particle because if two appeared on it at the same time, the capacitance energy would be increased and the gate voltage would have to be increased further to bring the forbidden states back into line. The device can also be made to conduct by applying a sufficiently large source-drain voltage so that the Fermi level of the source is lifted to the empty states of the nanoparticle with $N + 1$ electrons. This is not, however, the normal mode of operation for a transistor which we want to "switch on" only when the gate voltage is applied. In summary the gate voltage controls whether or not the nanoparticle conducts, one electron at a time, via quantum mechanical tunneling. It therefore acts as a single-electron FET; and if it can be realized, a whole new generation of electronics based on nanoparticles and quantum mechanics will be created.

ADVANCED READING BOX 5.3—ENERGY REQUIRED TO CHARGE A NANOPARTICLE

Capacitance is normally associated with the *mutual capacitance* of nearby conductors as in a capacitor in an electrical circuit, but every body that is electrically isolated from its environment has a self-capacitance, C, defined as the ratio of the charge placed on the object to its increase in electrical potential, V, that is,

$$C = \frac{Q}{V}.$$

The energy to charge a capacitor up with charge Q can be derived classically, assuming charge is continuous. Each element of charge, dq, put on the object requires an amount of work $dU = V dq$ to place it there, where V is the potential already acquired by the object. Since V is increasing as the body accumulates charge, to find the energy required to place a total amount of charge Q we need to evaluate the integral:

$$U = \int_0^Q V \, dq = \frac{1}{C} \int_0^Q q \, dq = \frac{Q^2}{2C}.$$

It appears from data gathered on SETs that this simple formula applies even when the charge is a single electron, for which

$$U = \frac{e^2}{2C}.$$

To evaluate this change in energy, we need to know the self-capacitance of a sphere. This can be derived from the potential of a point charge, which is (for a charge in vacuum)

$$V = \frac{Q}{4\pi \varepsilon_o r}.$$

It is easy to see that this potential distribution is the same as that of a sphere with charge Q outside the sphere. The equipotential surfaces of the point charge are spheres; thus if we create a conducting sphere with charge Q centered on the charge, it will have no effect on the field outside the sphere. Thus the capacitance of the sphere is

$$C = \frac{dQ}{dV} = 4\pi\varepsilon_0 r.$$

Thus the energy required to charge a spherical nanoparticle of diameter, d, with a single electron is

$$U = \frac{e^2}{4\pi\varepsilon_0 d}.$$

So for a particle with a diameter of 10 nm the energy required to place a single electron charge on it is 2.3×10^{-20} J or 0.144 eV. As discussed in the text, for an SET to work at room temperature, this has to be much greater than kT at 300 K or $\gg 0.025$ eV, which is the case for a 10-nm particle. The critical diameter for room temperature operation is $\sim$40 nm.

A critical parameter with regard to getting an SET to work is to make the capacitance small enough so that the energy shift in the electron states $e^2/2C$ is much greater than thermal energy; and for practical applications this needs to be much greater than the thermal energy at room temperature, which is about 0.025 eV/electron. Otherwise the tunneling can still be achieved with thermal assistance with the gate off, and the gate electrode would not control the operation of the device. Advanced Reading Box 5.3 shows that if the channel was a metal nanoparticle, this condition requires the particle to be smaller than $\sim$40 nm.

Structures showing room temperature SET behavior have been reported since 1995 [9], and since 1997 several groups have attempted to deliberately place single nanoparticles between electrodes with insulating gaps to directly set up the electrode structure shown in Fig. 5.11 [10–14]. A typical procedure for fabricating a nanoparticle SET is illustrated in Fig. 5.12 and is based on the method of Hong et al. [14]. The source and drain electrodes can be patterned using electron beam lithography (EBL; see Chapter 4, Section 4.2.1) on top of a very thin insulator on a conducting substrate, which forms the gate electrode (Fig. 5.12a). The tunneling barriers are synthesized by coating the electrode structure with a thin layer of insulator, and a particularly useful method is to use thiol molecules to produce a self-assembled monolayer (SAM; see Chapter 4, Section 4.4.5) as shown in Fig. 5.12b. This produces a uniform thickness of insulator (the length of the thiol molecule) and also guarantees that if a gold particle can be deposited into the electrode gap somehow, it will strongly bond there. The innovative method used by Hong et al. was to exploit a phenomenon

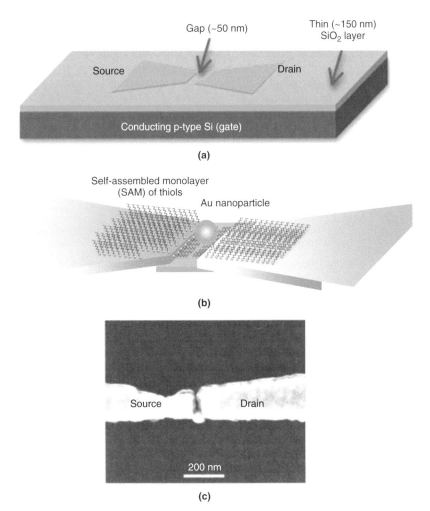

Fig. 5.12   Synthesis of single-nanoparticle SET. Synthesis of SET with the channel consisting of a single Au nanoparticle following the method of Hong et al. [14]. (a) The source and gate electrodes are patterned onto an oxide coated conducting Si substrate which forms the gate electrode. (b) The electrode structure is coated with a self-assembled monolayer (SAM) of thiols (1,8 octanedithiol in this case). This layer forms the insulating tunnel junctions. Au colloid is then deposited onto the electrodes and Au nanoparticles are attracted to the gap by dielectrophoresis. (c) SEM image of the successful placement of a single 50 nm Au nanoparticle at the gap. Reproduced with the permission of the American Vacuum Society from Hong et al. [14].

known as *dielectrophoresis*, which is a force exerted on an uncharged particle in a nonuniform electric field. It has been reported since the early 1950s [15] and has enjoyed a revival recently as its usefulness in guiding nanoparticles in suspension has been realized. In the study reported by Hong et al. it was

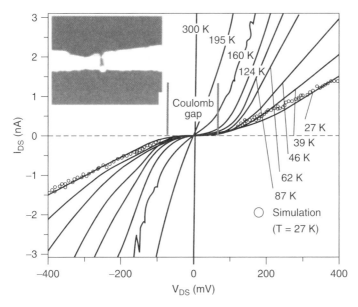

Fig. 5.13   Electronic behavior of single nanoparticle SET. Electronic behavior of the device shown in Fig. 5.12(c) for temperatures in the range 27 K to 300 K. At 27 K the gap due to Coulomb blockade is clear, that is, the device does not conduct until the source drain voltage increases beyond a threshold value. The size of the gap corresponds to Coulomb blockade by a Au nanoparticle with a size of 43 nm, which is close to that observed in the SEM image. The open circles show a simulation of the behavior assuming the basic theory outlined in Advanced Reading Box 5.3. Reproduced with the permission of the American Vacuum Society from Hong et al. [14].

exploited by depositing a tiny drop of a gold colloid onto the electrodes and then applying an ac electric field via the same electrodes to generate a force on the nanoparticles in suspension. The maximum field gradient occurs at the gap, and the gold nanoparticles naturally migrate there. They found that by varying the electric field parameters, they could control the number of nanoparticles gathered at the gap. Figure 5.12c shows a successful placement of a single 50-nm-diameter Au nanoparticle onto the SAM-coated electrodes. This is another nice example of combining bottom-up and top-down methods to construct nanostructures as discussed in Chapter 4, Section 4.3.

The current through the gold nanoparticle shown in Fig. 5.12c as a function of the source-drain voltage, V_{DS}, is displayed in Fig. 5.13 at temperatures ranging from 27 K to 300 K for zero gate voltage. At 27 K it is evident that the nanoparticle does not conduct over a range of V_{DS}, and this is the conductance gap due to Coulomb blockade. At 27 K the size of this gap corresponds to a nanoparticle self-capacitance of 2.3×10^{-18} farads, which in turn corresponds to a particle diameter of 43 nm (see Advanced Reading Box 5.3). The device is thus behaving according to the simple theory outlined above. With this particular device it proved problematic to control the behavior using the gate, but conductance oscillations as a function of the gate voltage were observed with more than one nanoparticle in the gap.

This is not a room temperature device but clearly demonstrates that Coulomb blockade transistors based on single nanoparticles can, at least in principle, be synthesized. By a slightly more hit-and-miss method, Luo et al. [12] succeeded in producing much smaller Au nanoparticles anchored by alkanedithiols to Au electrodes and demonstrated larger Coulomb gaps and room temperature switching behavior. Single-electron transistors have also been fabricated using C_{60} molecules, and these show up interesting additional effects in the electrical conduction behavior including the observation of a coupling between the motion of the molecule and electron hopping [16].

5.4 CARBON NANOELECTRONICS

The previous section presented the possibility of building an SET based on a C_{60} molecule, and there are also a host of suggested electronic devices based on fullerenes, carbon nanotubes, and even graphene sheets. Nanotubes can be semiconducting or conducting; thus, as well as acting as devices, they can be used as molecular-sized wiring to interconnect components in a circuit. This has led to speculations on future scale-ups in the density of electronic devices based largely around carbon (that is, carbon nanostructures) rather than silicon. The title of this section is "Carbon Nanoelectronics" to distinguish it from "Carbon Electronics," which traditionally means switching from silicon to diamond. When doped, diamond is, in some ways, a better semiconductor material than silicon, and a lot of research has been devoted to finding ways to produce thin diamond films onto which electronic circuits can be patterned.

An alternative to using a nanoparticle to construct an SET is to fabricate a device using a length of carbon nanotube. This was achieved in a metallic tube positioned between two Au electrodes by Postma et al. [17], who created the tunneling barriers along the tube by kinking it with an atomic force microscope (AFM; see Chapter 4, Section 4.4.4). Figure 5.14 shows their carbon nanotube SET at various stages of synthesis on top of an oxidized Si substrate that forms the gate electrode. The kink at each end of the short straight segment acts as a tunnel barrier, and the device behaves in a manner similar to that of the nanoparticle SETs described in the previous section. It showed a Coulomb gap of 120 meV and a conductance that could be controlled by the gate voltage.

One limitation of all the devices described so far is that they are individual transistors; and though it would be possible to fabricate many of them on the same substrate, this would mean that they all share a common gate electrode and would not be individually controllable. A step toward carbon nanotube integrated circuits was taken by Bachtold et al. [18] (from the same group that produced the carbon nanotube SET in Fig. 5.14), who reported the construction of circuits employing up to three carbon nanotube transistors, each with an independently

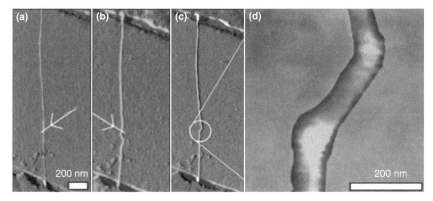

Fig. 5.14 **Synthesis of carbon nanotube SET.** Stages in the construction of a carbon nanotube SET using an AFM tip. The conducting nanotube is first placed between Au electrodes, which are the source and drain, on an oxidized Si substrate, which forms the gate electrode. The thin insulating tunnel junctions were produced by inducing buckles into the tube using the same AFM tip that was used to image the structures. (a) Initially straight nanotube showing position and direction of first AFM kink. (b) AFM image taken after first kink showing position and direction of second kink. (c) and (d) Final doubled kinked structure. Reproduced with the permission of the American Association for the Advancement of Science (AAAS) from H. W. Ch. Postma et al. [17].

controlled gate electrode. The nanotube transistors were not SETs relying on Coulomb blockade but instead employed semiconducting carbon nanotubes to fabricate a traditional FET architecture (see Fig. 5.9) but with a channel width of just one nanotube. These devices operate by the gate voltage being able to shift the Fermi level through the valence band, into the gap, and on into the conduction band. This gives control via the gate voltage of whether the tube conducts by holes (p-type) by electrons (n-type) or is nonconducting (off).

The nanotube integrated circuits were built in three steps. To begin with, EBL was used to deposit Al wires with a width ~ 1 μm, which formed the gate electrodes, onto the oxidized Si substrate. The insulating barrier between the gate electrode and the nanotube was the natural oxide that formed on the Al wires when they were exposed to air. Next, semiconducting nanotubes were deposited from a liquid suspension, and suitable individual nanotubes were maneuvered onto the gate electrodes using an AFM tip. Finally, after recording the positions of the nanotubes with respect to pre-patterned alignment marks, a suitable pattern of Au electrodes was deposited using EBL to form the source and drain electrodes and the interconnects for the circuit. Figure 5.15a shows a single nanotube FET on its gate electrode with source and drain deposited. Figure 5.15b shows a two-nanotube FET circuit with interconnects. The equivalent circuit is illustrated schematically in Fig. 5.15c and it is observed to be a three terminal device with one output and two input terminals. When power (-1.5 V) is applied to the device, assuming 0 V represents a logic "0" level and -1.5 V represents a logic "1" level, the digital outputs as a function of the digital inputs are shown in

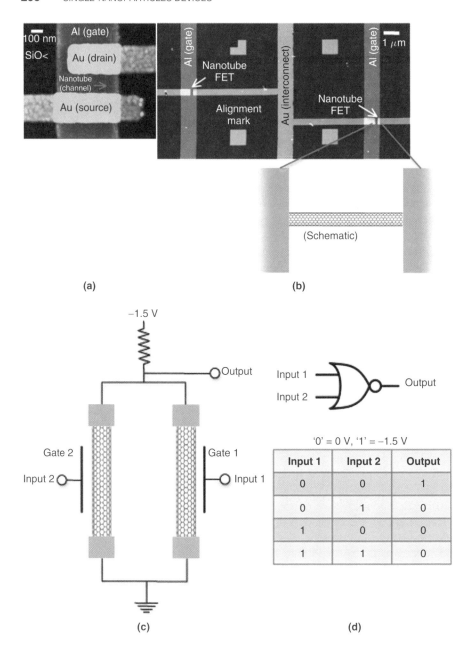

(a)

(b)

(Schematic)

−1.5 V

Output

Gate 2

Input 2

Gate 1

Input 1

(c)

Input 1
Input 2
Output

'0' = 0 V, '1' = −1.5 V

Input 1	Input 2	Output
0	0	1
0	1	0
1	0	0
1	1	0

(d)

the table in Fig. 5.15d. In digital electronics, such a table is known as a "truth table," and in this case it shows that the device gives an output of "0" unless both its inputs are also "0," which corresponds to the operation of a logic gate known as a NOR, whose circuit symbol is also shown. The NOR gate is one of the so-called "universal" gates because by wiring together NOR gates in the right configuration, it is possible to replicate every other type of logic gate. The carbon nanotube FET integrated circuit shown is therefore a building block of a microprocessor circuit of arbitrary complexity.

Building multiple CNT transistor circuits is a major step forward; but there is still one major hurdle to cross before the achievement of true nanoscale devices, and that is to provide nanoscale interconnects between the active electronic components. In the integrated circuit shown in Fig. 5.15b, although the individual nanotube FETs are tiny, they are several micrometers apart, which is not much different from existing Si-based integrated circuits. The source, drain, and gate electrodes and the interconnects fabricated by top-down methods can all be made smaller, but the ultimate step would be to wire the interconnects using conducting carbon nanotubes. Ultimately, this would all have to be achieved with some method in mind for scaling up and automating the process. At the moment the circuits rely on maneuvering individual nanotubes by AFM.

Semiconducting nanotubes can also be used to emit and detect light. In a conventional light-emitting diode (LED), p-type Si[1] is in intimate contact with n-type Si[2] to form a pn junction so that that electrons and holes are brought into close proximity. Initiating a current to flow through the device will cause electrons and holes to recombine and emit light as shown in Fig. 5.6c. Nanotube

---

Fig. 5.15 Construction of a carbon nanotube FET integrated circuit. (a) AFM image of an individual semiconducting carbon nanotube FET on its oxidized Al gate electrode and after deposition of the Au source and drain terminals. (b) AFM image of a two-nanotube FET integrated circuit after deposition of the Au electrodes and interconnects. The position of each nanotube in the circuit is indicated and the alignment shown in the schematic blow-up. Note that each nanotube FET has an independent gate electrode. (c) Schematic equivalent of the circuit in (b) showing that is can operate as a three terminal (two input and one output) device. (d) Assuming that logic '0' is represented by 0 V and logic '1' is represented by −1.5 V the table of output logic levels *vs*. input logic levels (the 'truth table') for the device is shown. It is observed to operate as a logic NOR gate (symbol shown), which is a Universal gate as NORs can be wired together to replicate the function of all other logic gates. The two nanotube FET integrated circuit is thus a building block for a microprocessor of arbitrary complexity. (a) and (b) reproduced with the permission of the American Association for the Advancement of Science (AAAS) from A. Bachtold et al. [18].

---

[1]p-type Si is doped with a trivalent metal such as Al and thus has an excess of holes in the valence band (see Fig. 5.6). Holes are the charge carriers when a current flows.
[2]n-type Si is doped with a pentavalent metal such as P and thus has an excess of electrons in the conduction band (see Fig. 5.6). Electrons are the charge carriers when a current flows.

FETs made from a semiconducting nanotube channel with suitable source and drain electrodes can be *ambipolar*; that is, the carriers of current in the tube can be both holes and electrons (Ref. 26, Chapter 3). For this type of device it is possible to control where along the nanotube electrons and holes can meet; thus it is not only possible to get a nanotube FET to act as a tiny LED, but it is possible to move the point of emission along the tube by changing the gate voltage. Coupled with the fact that there is a wide range of control over the bandgap of the tube (see Chapter 3, Section 3.10) and thus a wide range of wavelengths that the nanotube can emit, it is clear that carbon-based LEDs will become important devices once the synthesis problems have been overcome.

Semiconducting carbon nanotubes can also detect photons because an absorbed photon can create an electron–hole pair as in Fig. 5.6a, which will change the conductivity of the tube. This effect was first demonstrated by Freitag et al. in 2003 for individual carbon nanotubes incorporated into an ambipolar FET device structure [19]. An electric field applied between the source and drain electrodes separated the electron hole pairs as they were created and so a photocurrent was set up when the device was exposed to light. More recently, researchers at the Sandia National Laboratories in Livermore, California have developed nanotube photodetectors that detect light over a wide range of wavelengths using a different process that is related to way biological retinas work [20]. The tubes are coated with *chromophores*, which are molecules that change shape when excited by a particular wavelength of light. This change in shape results in a change in the electric dipole of the molecule that in turn changes the conductivity of the nanotube. The researchers succeeded in coating lines of nanotubes connected to electrodes with three different chromosphores that respond to red, green, and blue light, each having a different effect on the tube conductivity. This array can thus "see in color" and is a first step toward the synthesis of artificial retinas.

In this chapter, several distinct types of functionalised nanodevices, belonging under the general banner of evolutionary nanotechnology, have been described. These are: magnetic nanoparticles as memory elements, nanoparticle quantum dots, nanoparticle single-electron transistors (SETs), carbon nanotube SETs, and carbon nanotube field-effect transistors (FETs). An important class that has been omitted so far is bio-functionalized nanoparticles designed to interact with living tissue and perform some therapeutic or diagnostic function. Chapter 6 is devoted to this type of nanostructure.

PROBLEMS

1. In Richard Feynman's lecture "There's Plenty of Room at the Bottom," he evoked the possibility of writing the entire 40 million words of the *Encyclopaedia Brittanica* on the head of a pin. Assuming that the pin has a head diameter of 1 mm and that each letter is written using an 8-bit ASCII code with each bit being stored on a single magnetic nanoparticle, determine the particle diameter required to achieve the above data storage

density. Assume that the particles are in a square array and the gap between their circumferences has to be at least a particle radius to avoid significant magnetic interaction between them.

2. Calculate the blocking temperature of a 5-nm magnetic nanoparticle with a uniaxial magnetic anisotropy of 5×10^5 J/m^3 if it has to have a mean lifetime of 40 years.

3. If the maximum magnetic anisotropy that could be obtained in magnetic nanoparticles was 10^7 J/m^3, determine the fundamental limit of data storage density that could be achieved in a single-nanoparticle bit data storage medium with a lifetime of 40 years in Tb/in.^{2}. Assume that the particles are in a square array and the gap between their circumferences has to be at least a particle radius to avoid significant magnetic interaction between them.

4. The semiconductor CdSe has a bandgap of 1.74 eV. Quantum dots of CdSe passivated with ZnS are used to produce fluorescence of different QD colors. Determine the size of the CdSe core required to produce fluorescence at a wavelength of 400 nm (violet). Use the electron and hole effective masses given in Advanced Reading Box 5.2.

5. Determine the minimum particle diameter of a gold nanoparticle to work as a single-electron transistor using Coulomb blockade at room temperature.

REFERENCES

1. C. Binns, M. J. Maher, Q. A. Pankhurst, D. Kechrakos, and K. N. Trohidou, Magnetic behavior of nanostructured films assembled from preformed Fe clusters embedded in Ag, *Phys. Rev. B* **66** (2002), 184413.

2. S. Sun, Recent advances in chemical synthesis self-assembly, and applications of FePt nanoparticles, *Adv. Mater.* **18** (2006), 393–403.

3. M. R. Koblischka, U. Hartmann, and T. Sulzbach, Improvements of the lateral resolution of the MFM technique, *Thin Solid Films* **428** (2003), 93–97.

4. V. L. Mironov, B. A. Gribkov, S. N. Vdovichev, S. A. Gusev, A. A. Fraerman, O. L. Ermolaeva, A. B. Shubin, A. M. Alexeev, P. A. Zhdan, and C. Binns, Magnetic force microscope tip induced remagnetization of CoPt nanodiscs with perpendicular anisotropy, *J. Appl. Phys.* **106** (2009), 053911 (8pp).

5. H. J. Mamin, B. D. Terris, L. S. Fan, S. Hoen, R. C. Barrett, and D. Rugar, High-density data storage using proximal probe techniques, *IBM J. Res. Dev.* **39** (1995), 681–700.

6. S. C. Minne, S. R. Manalis, A. Atalar, and C. F. Quate, Independent parallel lithography using the atomic force microscope, *J. Vacuum Sci. Technol. B* **14** (1996), 2456.

7. M. Lutwyche, C. Andreoli, G. Binnig, J. Brugger, U. Drechsler, W. Häberle, H. Rohrer, H. Rothuizen, P. Vettiger, G. Yaralioglu, and C. Quate, 5×5 2D AFM cantilever arrays a first step towards a terabit storage device, *Sens. Actuators* **73** (1999), 89–94.

8. http://domino.watson.ibm.com/comm/pr.nsf/pages/rsc.millipede.html

9. Y. Takahashi, M. Nagase, H. Namatsu, K. Kurihara, K. Iwdate, Y. Nakajima, S. Horiguchi, K. Murase, and M. Tabe, Fabrication technique for Si single-electron transistor operating at room temperature, *Electron. Lett.* **31** (1995), 136–137.

10. D. L. Klein, R. Roth, A. K. L. Lim, A. P. Alivisatos, and P. L. McEuen, A single-electron transistor made from a cadmium selenide nanocrystal, *Nature* **389** (1997), 699–701.

11. S. H. M. Persson, L. Olofsson, L. Gunnarsson, and E. Olsson, Self-assembled single electron tunneling devices with organic tunnel barriers, *Nanostructured Mater.* **12** (1999), 821–824.

12. K. Luo, D. —H. Chae, and Z. Yao, Room-temperature single-electron transistors using alkanedithiols, *Nanotechnology* **18** (2007), 465203.

13. Y. Yang, Masayuki, and M. Nogami, Room temperature single electron transistor with two-dimensional array of $Au-SiO2$ core–shell nanoparticles, *Sci. Technol. Adv. Mater.* **6** (2005), 71–75.

14. S. H. Hong, H. K. Kim, K. H. Cho, S. W. Hwang, J. S. Hwang, and D. Ahn, Fabrication of single electron transistors with molecular tunnel barriers using ac dielectrophoresis technique, *J. Vac. Sci. Technol. B* **24** (2006), 136–138.

15. H. A. Pohl, The Motion and precipitation of suspensoids in divergent electric fields, *J. Appl. Phys.* **22** (1951), 869–871.

16. H. Park, J. Park, A. K. L. Lim, E. H. Anderson, A. P. Alivisatos, and P. L. McEuen, Nanomechanical oscillations in a single-C_{60} transistor, *Nature* **407** (2000), 57–60.

17. H. W. C. Postma, T. Teepen, Z. Yao, M. Grifoni, and C. Dekker, Carbon nanotube single-electron transistors at room temperature, *Science* **293** (2001) 76–79.

18. A. Bachtold, P. Hadley, T. Nakanishi, and C. Dekker, Logic circuits with carbon nanotube transistors, *Science* **294** (2001), 1317–1319.

19. M. Freitag, Y. Martin, J. A. Misewich, R. Martel, and P. Avouris, Photoconductivity of single carbon nanotubes, *Nano Lett.* **3** (2003), 1067–1071.

20. X. J. Zhou, T. Zifer, B. M. Wong, K. L. Krafcik, F. Léonard, and A. L. Vance, Color detection using chromophore-nanotube hybrid devices, *Nano Lett.* **9** (2009), 1028–1033.

CHAPTER 6

Magic Beacons and Magic Bullets: The Medical Applications of Functional Nanoparticles

Nanotechnology is a multidisciplinary activity, breaking through the barriers between physics, chemistry, and biology. Since its emergence there have been attempts to exploit it in medicine and to explore whether it is possible to use it to overcome certain persistent problems in conventional medical treatments. An example is in cancer therapies that use highly toxic drugs, and the treatment is a balancing act between harming the patient and treating the cancer. There has been a steady improvement in the 5-year survival rate, which in the United States averaged over all cancers in the last 10 years has reached 66% [1]. The basic abiding problem is that it is hard to concentrate the available drugs in the tumour tissue so that doses are limited by the toxicity of the drug. Improvements in diagnosis and early detection are also very important and in both therapy and diagnosis, nanoparticle-based techniques are being intensively researched.

Nanoparticles are already available for some applications—for example, as magnetic resonance imaging (MRI) contrast enhancers (see Section 6.3.1). This is only the beginning, however, and it is believed that nanotechnology will make a significant contribution to improvements in health care. Most envisaged future medical applications use nanoparticles that are programmed to individually perform a function, thus conforming to the definition of evolutionary nanotechnology given in the introduction, and this chapter is devoted to these. There are also many ideas for rapid diagnosis and screening based on top-down nanotechnology—for example, the diagnostic arrays synthesized using dip-pen nanolithography (DPN) described in Chapter 4, Section 4.4.5—but these will not be covered in this chapter. Looking further into the future, there is the possibility of using nanotechnology to build devices that replicate functions of the body—for example, the artificial retina described in Chapter 5, Section 5.4. Finally, there are speculations on the use of radical nanotechnology to provide diagnosis and effect repairs at the cellular level. These are discussed in Chapter 7, Section 7.3. Here the focus is on diagnosis and treatment of disease using nanoparticles. This chapter,

Introduction to Nanoscience and Nanotechnology, by Chris Binns
Copyright © 2010 John Wiley & Sons, Inc.

more than any other, emphasizes the multidisciplinary nature of nanotechnology encompassing physics, chemistry, biology, engineering, and medicine.

6.1 NANOPARTICLES INTERACTING WITH LIVING ORGANISMS

6.1.1 Targeted Nanovectors for Therapy and Diagnosis

Generally, in nanoparticle-based medical applications the particles are programmed to locate specific types of cells (usually tumor cells) and then either perform some action to kill the cell (magic bullets) or show up with a very high sensitivity using an external diagnostic probe such as magnetic resonance (MRI) or ultrasound imaging (magic beacon). The magic bullet idea goes back to the physician Paul Ehrlich (1854–1915), who first coined the phrase and who received the 1908 Nobel Prize in Medicine. Ehrlich's original idea was to deliver drugs to specific organisms, and using nanoparticles this is certainly possible. As discussed below, however, in some cases one can dispense with the drug and use the nanoparticle itself to provide the therapy by heating up in response to an external stimulus and killing the cell. This is known as hyperthermia (Section 6.2).

A fully functional nanoparticle programmed to seek and kill diseased cells, often called a nanovector, is shown schematically in Fig. 6.1a. It consists of a core particle, which can be a passive carrier of drugs, such as a hollow spherical liposome (Fig. 6.2) or a solid particle that will perform some function in response to an external stimulus. In its most versatile form the nanovector will have a number of other functionalities built in. These can be permeation enhancers that help it penetrate the walls of blood vessels in the region of the tumor and reach the tumor cells themselves and contrast enhancers that boost its signal in diagnostic MRI or in optical imaging. On its surface the particle can be coated with targeting molecules or moieties that will specifically bind to receptors on tumor cells. The types of targeting molecules that can be used are described in Section 6.1.5. The surface of the particle can also be coated with polyethylene glycol (PEG) molecules that prevent the nanovector from being taken up by macrophages, which form a line of defense in the body's innate immune system.

If the nanovector is a drug carrier, it needs to release its drugs when it reaches the target cell. This can be done by having the core particle made of a polymer that erodes away at a predetermined rate emptying its contents, or, in the case of a targeted liposome, the core particle will merge with the membrane of the target cell, thus delivering its cargo. One can of course attach the targeting molecules directly to the molecules of the drug so that they are delivered to where they are required, but the advantage of the nanovectors is that they can deliver large amounts of therapeutic or imaging agents for each successful targeting event.

Many simpler types of nanovector are also under study. For example, Fig. 6.1b shows a small ($\sim$10 nm) magnetic nanoparticle fitted with targeting moieties. This is also programmed to find a specific type of cell; and when in place, it

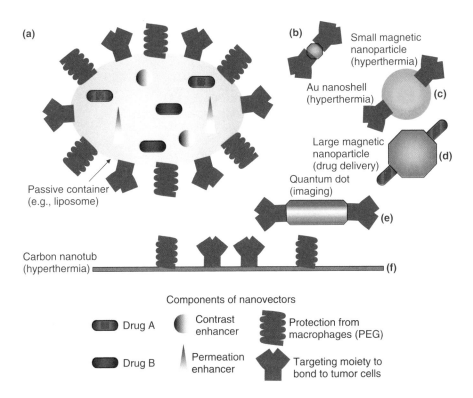

Fig. 6.1   Generic types of nanovectors suggested for diagnosis and treatment of cancer. (a) Fully functional versatile nanovector that consists of a container (~100 nm across) of drugs, contrast enhancers for MRI or other imaging methods, permeation enhancers that help it pass through blood vessel walls to reach the target cell and drugs that can be released at the target site to kill the diseased cell. The particle has targeting moieties on the outside so it will only attach to the diseased cell and protection (PEG molecules in this case) from macrophages, which are part of the body's passive immune system. (b) Simpler nanovector consisting of a small (~10 nm) nanoparticle that is targeted to the right type of cell and is then heated by an external oscillating magnetic field applied from outside the body. Sufficient numbers of these can generate enough heat locally to kill the diseased cell (Section 6.2.2). These can also act as their own contrast enhancers in MRI images. (c) Au nanoshell, which is a powerful infrared absorber, with targeting moieties used for optical hyperthermia in which the local heat is generated by low power infrared radiation applied from outside the body (Section 6.2.3). (d) Drugs attached to large (~100 nm) magnetic nanoparticles that are concentrated in the region of the tumor by a strong external magnetic field gradient. (e) Targeted quantum dots that emit at a suitable wavelength at their target site can be used to sensitively image tumors. (f) Targeted nanotubes, which are powerful infrared absorbers and may be used in hyperthermia similarly to Au nanoshells (Section 6.2.4).

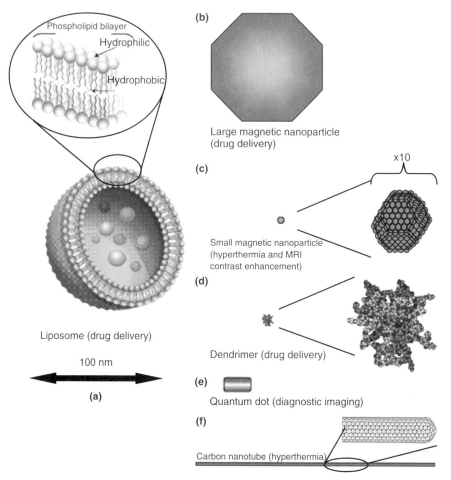

Fig. 6.2 **Types of core nanoparticles used in medical applications (to scale).** (a) Liposome (~100 nm across) able to carry and deliver drugs (reproduced with permission from Jane Wang (artist) fromhttp://www.bioteach.ubc.ca/Bio-industry/Inex/). (b) Large magnetic nanoparticle used to deliver attached drugs to specific locations controlled by an external magnetic field (Section 6.1.6). (c) Small magnetic nanoparticle (superparamagnetic) used for magnetic hyperthermia (Section 6.2.2) or MRI contrast enhancement (Section 6.3.1). (d) Dendrimer particle used to carry and deliver drugs (reproduced with permission from dendritech inc. http://www.dendritech.com/pamam.html). (e) Quantum dot used for diagnostic imaging. (f) Carbon nanotube used for optical hyperthermia.

can be heated by an oscillating magnetic field applied from outside the body, thus dumping heat into the local cell (hyperthermia; see Section 6.2.2). Relatively small temperature rises of the order of 10°C are required to kill cells, so this can be an effective therapeutic method and dispenses with the need for the highly toxic drugs. The same nanoparticle will also boost the signal in an

MRI scanner so that it acts as its own contrast enhancer for diagnosis. A similar hyperthermic approach uses Au nanoshells (Fig. 6.1c), which are very effective absorbers of infrared light. Once the nanovectors are in place, the light can be applied at an intensity that is too low to damage normal tissue, but the diseased cells with the attached absorbers will heat preferentially and be killed (Section 6.2.3). Figure 6.1d shows a nanovector consisting of a large (~100 nm) magnetic nanoparticle that has the drugs attached on the outside and doesn't use targeting molecules but is instead steered into the region of the tumor by a strong external magnetic field gradient (Section 6.1.6). After concentrating the magnetic nanoparticles in the right place, they must be persuaded to release their drugs; one possible approach is to attach the drugs with a linker that releases them in response to a small temperature rise. Applying a low power level of heating with an external oscillating magnetic field as used for hyperthermia will effect the release. Figure 6.1e shows a targeted quantum dot that can be "lit up" (as discussed in Chapter 5, Section 5.2) at its target location and be imaged externally for diagnosis (Section 6.3.3). Carbon nanotubes (Fig. 6.1f) described in Chapter 3 have recently been studied as nanovectors. They are strong optical absorbers of infrared light and could be used for optical hyperthermia in a manner similar to that of Au nanoshells (Section 6.2.4). In addition, their uptake in the body is different from that of other nanomaterials (Section 6.1.4).

In some cases the targeting can be passive, dispensing with the need for targeting moieties by using the known unique properties of the tumor environment to concentrate the nanoparticles. For example, there is inefficient drainage from the region of a tumor due to a damaged lymphatic system leading to fluid retention. Another effect, known as enhanced permeability and retention (EPR), is due to the fact that the blood vessels around a tumor site are leaky and nanoparticles tend to get pushed out of the blood vessels and thus accumulate in the region. Due to these effects, nanoparticles injected into the environment of a tumor will tend to stay there and a suitably designed polymer capsule will erode away, releasing the drug in a controlled way over a period of time. The efficiency of the retention of the nano-capsules depends on their size and in general they need to be around 100 nm across. Some treatments in this form are already available, for example, liposome-encapsulated molecules of a drug called doxorubicin were approved 10 years ago for the treatment of Karposi's sarcoma and are now used to treat breast and ovarian cancers.

Clearly with the more advanced and targeted nanovectors described above, treatments will become more effective. Several thousand different nanovectors have been reported in the literature, though only a tiny minority have been tested for their effectiveness in cancer treatments. The most advanced method is magnetic nanoparticle hyperthermia, which has progressed to Phase II clinical trials (Section 6.2.2), though so far without biological targeting. The following sections describe the details of targeting, hyperthermia, magnetic delivery, and contrast enhancement.

6.1.2 Types of Core Nanoparticle in Nanovectors

The core nanoparticles onto which drugs, targeting moieties, and so on, are attached fall into several generic classes depending on the application. Figure 6.2 shows the main types, approximately to scale, stripped of their functional agents.

Liposomes (Fig. 6.2a) have been used for more than a decade for drug delivery and have found their way into a host of applications including cosmetics where they are used to deliver anti-aging, skin-whitening, and moisturizing agents beneath the skin. They consist of a basic biological membrane rolled into a sphere that encloses the liquid in which it was formed. Membranes are prolific in biology, forming the outer skin of all cells and enclosing structures within cells. The basic material of a typical membrane is made from a double layer of phospholipid molecules (inset of Fig. 6.2a), which consist of a hydrophilic and a hydrophobic part. In an aqueous environment the phospholipid molecules bind together to form a carpet-like layer and then the layers will come together to isolate the hydrophobic parts from the liquid. This double layer forms a very effective watertight seal and in real biological membranes contains a host of other agents that allow it to pass material either way in a controlled fashion. The basic raw material is the same, however; and when liposomes meet biological cells, they tend to unravel and merge with the membrane of the cell, thus releasing their contents. The cargo to be carried by the liposome can be simply loaded by including the drugs and other agents in the liquid in which the liposome is formed. In addition, it is possible to dissolve certain types of drugs within the membrane itself.

Figure 6.2b shows the typical size of magnetic nanoparticle that can be used to steer drugs to the right region by externally applied magnetic fields. For reasons discussed in Section 6.1.6, a sufficient force can only be applied if the particle is larger than about 50 nm. The material used in the particles is usually iron oxide [maghemite (Fe_2O_3) or magnetite (Fe_3O_4)]. Often they are coated with dextran—a sugar to improve their biocompatibility and to make it relatively easy to attach other agents such as targeting moieties and drugs to the surface. Although the oxides are relatively easy to prepare, they have a saturation magnetization of only about 25% of that of pure iron and are much weaker magnets. This puts an enormous demand on the strength of the applied magnetic field gradient and limits the method to targeting near the surface of the body. Although pure iron nanoparticles would be much more effective, they are also more toxic than the oxides and cannot be used unless they are coated with a biocompatible material. Recently, a process has been developed that produces core-shell particles with a pure iron core coated in an iron oxide shell [2]. These will experience a much greater force enhancing the versatility of magnetic targeting.

Figure 6.2c shows the typical size of magnetic particle used for hyperthermia. For reasons discussed in Section 6.2.2 the particles produce the most effective heating in an applied oscillating magnetic field if they are about 5 nm in size. Again most research carried out so far has used iron oxide nanoparticles, but heating rates could be dramatically improved if pure iron or cobalt nanoparticles could be used after they had been rendered nontoxic by coating them in a biocompatible material. The coating would also facilitate the bonding of targeting moieties used to attach the nanoparticles to diseased cells. The same size of particle is used to enhance the contrast in MRI images (Section 6.3.1) and dextran-coated iron oxide nanoparticles are licensed and commercially available in injectable suspensions for this use.

A new type of drug-carrying nanoparticle that has attracted a lot of attention recently is known as a dendrimer (Fig. 6.2d). These were described in Chapter 4, Section 4.1.7 and are polymers that are grown in a tree-like structure from a central core in "generations." Each generation added increases the size of the particle and also the number of active sites on the surface that can be used for bonding external agents. For example, a five-generation (or G5) dendrimer has a diameter of 5.4 nm and 128 active sites on its surface, while for a G6 polymer these values increase to 6.7 nm and 256, respectively. Dendrimers up to G10 are commercially available.

Figure 6.2e shows the approximate size of a quantum dot that could be used for diagnostic imaging. Quantum dots were discussed in Chapter 5, Section 5.2, and it was shown that they could be made to emit in a range of colors depending on the dot size. Tissue is fairly transparent to certain wavelengths so that targeted quantum dots can be attached to the desired cells and the location and distribution of these cells can be imaged from outside the body (Section 6.3.3).

Finally carbon nanotubes, which have been the focus of intense research in materials science have recently been investigated as nanovector carriers. They have a high optical absorbance at infrared frequencies and could be used for optical hyperthermia in a similar manner to Au nanoshells (Section 6.2.4). They can be made water-soluble by attaching PEG to them and it has been shown that in this state they traverse through the membranes and into the interior of cells.

6.1.3 Some Elementary Human Cell Biology

Most of the proposed ideas put forward in medical nanotechnology involve nanoparticles interacting with cells. A broad overview of living cells is thus helpful here. Human cells come in a variety of shapes and sizes, but generally they are part of the "micro" world rather than the nanoworld, being typically several microns across and thus about a thousand times larger than nanoparticles. The largest cell, the human ovum, is huge on the nanoscale, being typically 100 μm (0.1 mm) across and thus just about visible to the naked eye. Figure 6.3 shows two of the smallest cells in the body, with sizes less than 10 μm—that is, a leukocyte or white blood cell (Fig. 6.3a) and an erythrocyte or red blood cell (Fig. 6.3b). There are several million of these per cubic millimeter of our blood. The inset in the red blood cell image shows an electron microscope image of the typical composition of blood with red and white blood cells and platelets. Nanoparticles that could be used for therapy and diagnosis are shown to scale in Figs. 6.3c and 6.3d. Broadly, the size-dependent properties of nanoparticles

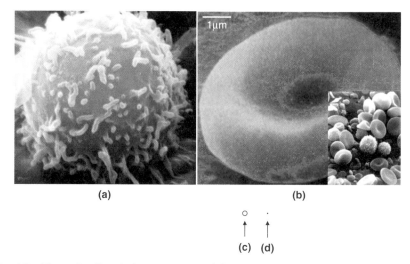

(a) (b)

o ·
↑ ↑
(c) (d)

Fig. 6.3 Sizes of cells relative to nanoparticles. (a) Leukocyte or white blood cell (b) Erythrocyte or red blood cell. There are millions of these per cubic millimeter of our blood. The inset in (b) shows the typical constitution of blood with red and white blood cells and platelets. These are among the smallest cells found in the body with diameters less than 10 μm. (c) To scale a 100 nm particle (e.g. liposome, magnetic nanoparticles for drug delivery) (d) a 5 nm nanoparticle (e.g. dendrimer, magnetic nanoparticle for hyperthermia, quantum dot).

define two size ranges that are useful in medical applications (Fig. 6.2)—that is, "large" nanoparticles with sizes in the range 50–100 nm diameter and "small" nanoparticles in the size range 2–10 nm. It is evident from Fig. 6.3 that even the largest therapeutic nanoparticle has an insignificant volume relative to the smallest human cell.

Each cell is alive and as long as it is kept supplied with nutrients will continue to perform its designated function within the body. The internals of a typical cell, shown in Fig. 6.4, contain an array of chemical processors ("organelles") that enable the cell to distribute nutrients, extract energy from them, and perform other necessary functions. In some ways it is a miniature analog of a living animal with the organelles taking the role of organs. All the organelles are within a thick jelly-like substance called the cytoplasm, which is contained within a flexible membrane that is akin to human skin in that it discriminately allows substances to pass into or out of the cell. The process where materials pass from the environment through the membrane into the cell interior is known as *endocytosis*. Within the cell is a bone-like structure called the cytoskeleton, which maintains the cell shape and also allows it to change shape. All the genetic information that passes on the instructions to provide a new copy of the cell is in the DNA, which is contained within the cell nucleus, usually inside a second membrane. In medical applications of nanoparticles, we want the particle to somehow recognize a diseased cell, attach to the membrane or pass through it,

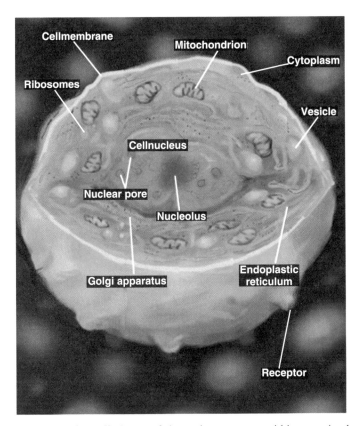

Fig. 6.4   Anatomy of a cell. Some of the main structures within an animal cell. The outer skin is the *membrane*, which is filled with the jelly-like *cytoplasm* and the internal structures (*organelles*) that allow the cell to function. These include the *mitochondria* that produce chemical energy from the nutrients, *ribosomes* that assemble proteins, the *golgi apparatus* that packages large molecules produced by other organelles and distributes them to the correct part of the cell, the *endoplasmic reticulum* that has several functions including the folding of proteins and vesicles that store or digest waste. At the heart of the cell is the *nucleus* containing DNA. The nucleus is itself an organelle that provides services for the cell including the production and assembly of ribosome components in the *nucleolus*. One communication mechanism of the cell with the outside world is via receptors on the cell membrane that are specialized to bond to specific molecules. The *targeted nanoparticle* therapies use these receptors to bond functionalized nanoparticles to tumor cells. The particles either carry drugs that are released at the destination cell or they can be made to heat up via an external stimulus to kill the cell. Image supplied courtesy of Abby Marsh and Bob Gross, Dartmouth University, USA.

and then release its load of attached therapeutic drugs or heat up in response to an external stimulus in order to kill the cell. It is important to realize that only small temperature rises are required. A temperature of 60°C will kill a cell instantly, while a temperature of 42°C sustained over a period will also kill it. Some of the more exotic proposals based on radical nanotechnology discuss the possibility of repairing the damaged cell (see Chapter 7, Section 7.3).

6.1.4 Uptake of Nanomaterials by the Body

Nanoparticles entering the body trigger a response from the *reticuloendothelial system* (RES), which comprises a host of cells distributed throughout the body and specialized to ingest, process, and/or transport foreign invaders to the liver and kidneys to be ejected in the feces and urine. The system has evolved to deal with a range of intruders including bacteria, viruses, toxins, and passive particles. In the discussion below we will assume that the therapeutic nanoparticles are passive (i.e., nontoxic) and that they enter the bloodstream directly (say by intravenous injection). For all the particles considered in this chapter it is safe to assume that even the "large" ones (~100 nm) are too small to block even the narrowest of blood vessels (capillaries), which have diameters of at least 5 μm.

Once in the bloodstream, the particles encounter leukocytes or white blood cells (Fig. 6.3a), which make up about 1% of whole blood by volume. These cells are designed to engulf foreign particles by a process known as *phagocytosis*. It is difficult for this to happen directly because generally both the leukocytes and particles have a negative charge (*zeta potential*) and tend to repel each other. This is overcome by an intermediate process (*opsonization*) in which circulating proteins, mostly antibodies (Section 6.1.5), bind to the particle. The proteins bound to the particle are attracted to receptors expressed on the leukocyte surface. Once attached, the membrane of the leukocyte surrounds the particle and brings it into the cell wrapped in a piece of membrane. The package consisting of the particle surrounded by cytoplasm and wrapped in membrane is referred to as a *phagosome*. The phagosome is transported through the cell to organelles (known as *lyosomes*) that process the particle further and attempt to break it down to basic constituents. For nanovectors that include biological components, these will be broken down by the lyosomes, but solid metal or oxide nanoparticles will be kept internalized and transported to the liver, spleen, and bone marrow, where specialist cells capture the nanoparticles and attempt to break them down further or get them ready for expulsion from the body. For example, if they end up in the liver, they are incorporated into the bile and removed by feces. Alternatively, depending on the size, they can be filtered by the kidneys and removed in the urine. The RES also consists of cells capable of phagocytosis dispersed in the tissues, which can remove debris and particles by other mechanisms. An example is the *mucociliary escalator* (described in Chapter 2, Section 2.2), which removes unwanted particles from the lungs.

In general, the residence time of particles within the bloodstream increases as the particle size decreases but can still be quite short. For a small (~10 nm) nanoparticle the blood residence time is typically hours and for large nanoparticles (~ 100 nm) it can be a few minutes. This can present a significant problem for treatment since nanoparticles with a very high performance *in vitro* are useless if *in vivo* they are removed before their targeting can work or before they can be activated. Therefore it is important to develop "stealth" technologies for the particles that prevent them being recognized by leukocytes, and this means preventing the initial step of opsonization. Clearly, such a thing is possible since the red blood cells (erythrocytes; see Fig. 6.3) evade the leukocytes and have a life span of about 120 days while they fulfill their task of transporting oxygen. One factor in determining the relatively long life span of erythrocytes is a coating of sugar polymers that prevent the adsorption of opsonins and the resulting recognition by leukocytes and macrophages. Similar coatings can be used to protect therapeutic nanoparticles.

A common artificial coating to evade the cells of the RES is polyethylene glycol (PEG; See Fig. 6.1). The mechanism by which PEG helps to protect therapeutic nanoparticles is still uncertain, but recent findings [3] suggest that PEG controls the buildup of charge that can occur on a nanoparticle surface as proteins accumulate. It appears that PEG can keep changes in nanoparticles surface charge to a minimum, which helps to hide them from the innate immune system. In general, to maximize the lifetime within the blood plasma of an artificially introduced particle, it should be as small as possible, be electrically neutral, and have a hydrophilic surface.

6.1.5 Biological Targeting

Whether one uses the nanoparticles as delivery vehicles of drugs or as therapeutic agents in themselves, their effectiveness is greatly enhanced if targeting moieties are attached to their surface, causing them to bind selectively just to tumor cells. There are a number of possible generic types of targeting molecules including antibodies, aptamers, peptides, and folate (folic acid). These are briefly described below.

Antibodies are naturally produced in our bodies as part of the adaptive immune system that identifies invading organisms and also "remembers" characteristics of invaders so that if they challenge the immune system again a more efficient response is evoked. An antibody is a large "y"-shaped molecule with a common base and variable ends that can bind to specific foreign antigens—that is, specific molecular structures on the surface of a foreign body. It is a bit like a socket set, where the handle is common but different sockets with different sizes and shapes can be attached on the end. The structure is shown in Fig. 6.5 and consists of two identical "heavy chains" and two identical "light chains." It is the ends of the two sets of chains that form the specialized sockets or *paratopes* that will efficiently bond only onto specific antigens. It is a big molecule with a size of tens of nanometers containing $\sim 20,000$ atoms.

In a healthy organism, specialized cells called B cells carry within their population every antibody that can be manufactured by the body. If a foreign invader

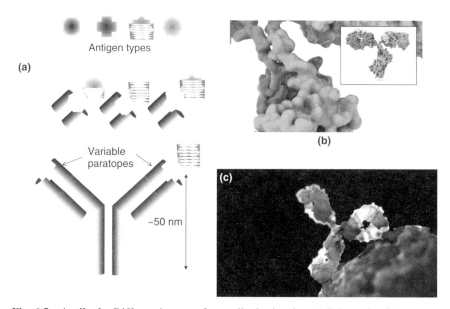

Fig. 6.5   Antibody. Different images of an antibody showing (a) Schematic of the 'socket set' view of the molecule indicating the variable end pieces or *paratopes* in mauve that can be replaced in order to bind efficiently onto different antigens. (b) Morphology of an antibody showing in detail the neck region and the two different chains. (c) An antibody attached onto an invading rhinovirus (common cold virus). Reproduced with permission from http://www.anatomyblue.com/.

is encountered, a B cell that binds successfully via its antibody to the alien starts a process that ends up with that cell and its progeny reproducing so that millions of copies of B cells with that specific antibody are produced. These will then seek other invading cells that express the same antigen. Normally the antibody-bearing B cells do not kill the invaders directly but mark them for destruction by other cells of the immune system.

The important point is that in a cancer, which is a colony of malfunctioning cells that is treated by the immune system as a foreign invader, the cells express antigens peculiar to that specific cancer that will bind to antibodies. Thus in types of cancer where specific antigen–antibody combinations are known, the antibody can be manufactured and attached to a nanoparticle, be it a magnetic nanoparticle, a quantum dot, or a drug-carrying vessel. If these nanoparticles are introduced into the area where the cancer resides they will target and bond to the tumor cells in preference to healthy cells. Producing specific antibodies is a complex process that starts with their manufacture in an animal that has been exposed to the cancer to be treated. The initial antibodies can be cloned but then have to be "humanized" so that they don't produce an immune response from the patient.

Antibody targeting of receptors on tumor cells is already in use where antibodies are introduced to help the body's immune response to specific cancers. An example is trastuzumab (Herceptin®), which binds to so-called "HER2" receptors on certain types of breast cancer. Antibodies are presently being tested as carriers of chemotherapeutic drugs where each antibody carries about 10 drug molecules. Attaching antibodies to nanoparticle carriers of drugs enables much bigger loads to be released at each successfully targeted site. Also, it is possible to attach antibodies to magnetic nanoparticles that will seek and bind to the target tumor cells. These will then show up with high sensitivity in an MRI image (Section 6.3.1); and they can also be used to kill the cell using hyperthermia (Section 6.2.2), avoiding the need for chemotherapeutic drugs.

Antibodies are the oldest and most studied type of targeting molecules and, having evolved over millions of years, are supremely good at their job—that is, identifying and binding to antigens—but they do have some disadvantages as targeting molecules for nanoparticles. One is just the sheer size of a complete antibody—typically tens of nanometers—which prevents it from passing through the cell membrane should that be required. They are also not suitable for treating diseases in the brain since they do not pass easily through the blood–brain barrier, again because of their large size. In addition, they are expensive to produce and a new population has to be started in an animal exposed to the pathogen that presents the required antigen, which means that antibodies cannot be prepared for very toxic pathogens. Finally, they have a well-developed system for communicating with the immune system and can often provoke an immune response that attacks the medicine.

This has led to the search for fragments of antibodies that will also target antigens—for example, just one of the arms of the "y." Such reduced molecules, about 10 times smaller than full-blown antibodies, are often referred to as *nanobodies*. They are just as good at bonding to antigens, but they cannot perform the full function of an antibody including triggering other processes within the immune system, which in this case is an advantage.

Another possible biological targeting system is based on DNA. This consists of *nucleobases*, which are the small molecules shown in Fig. 6.6 that, when combined with a sugar and phosphate group, can polymerize to form a phosphate "backbone" (Fig. 6.6) with the amino acid bases strung in a specific sequence along the chain. The sequence of the four bases labeled "A" (adenine), "T" (thymine), "C" (cytosine), and "G" (guanine) along the backbone will fit only onto another string of bases on another strand that are in a complementary sequence, like a lock and key. An aptamer is an artificially produced short section of DNA typically between 15 and 40 nucleotides long. When in solution the interactions between the nucleotides in the chain folds the aptamer into a complex three-dimensional shape, which, if optimized to a negative image of the shape of a receptor in the antigen, will bind tightly to it. This lock-and-key functionality is encapsulated in the name aptamer, which is derived from the Latin *aptus* (to fit). Since there are an enormous number of possible nucleotide sequences in even a short aptamer strand, there is a huge diversity of shapes and it should be possible to find suitable aptamers to target most antigens.

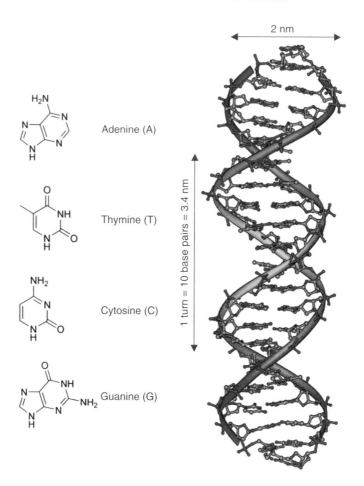

Fig. 6.6 Nucleobases and DNA. Left: The four nucleobases in DNA. Right: Typical section of DNA showing about 20 base pairs. The nucleobases are attached to separate sugar-phosphate backbones and the strands come together so that bases bond to form base pairs. There are 10 base pairs per turn of the strands and each base pair takes up a length of 3.4 nm on the molecule. DNA molecules can contain up to 2000 base pairs (~700 nm in length). An aptamer is an artificially produced short section of DNA typically between 15 and 40 nucleotides long. Pictures from Wikipedia page http://en.wikipedia.org/wiki/Dna.

Apatamers have significant advantages over antibodies as targeting molecules. They can be produced much more cheaply *in vitro* without involving the immune system of an animal, and they are much smaller than antibodies and even smaller than nanobodies. In addition, because they have none of the machinery of an antibody that communicates with the immune system, they invoke no immune response themselves. Figure 6.7 shows an *in vitro* image of prostate cancer cells, indicated by their nuclei (stained blue) and their cytoskeletons (stained green) taken by a team at the Brigham and Women's Hospital, Harvard. The

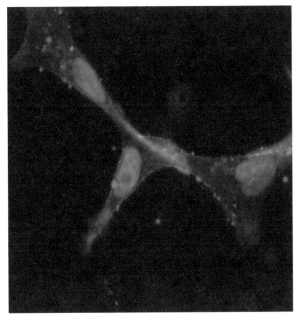

Fig. 6.7 Targeting using aptamers. Image of three prostate cancer cells that have taken up fluorescently labelled nanoparticles (shown in red) using an aptamer-targeting molecule. The cells' nuclei and cytoskeletons are stained blue and green, respectively. Such targeted nanoparticles are capable of getting inside cancer cells and releasing lethal doses of chemotherapeutic drugs to eradicate tumors. Reproduced with permission from Prof. Robert Langer, MIT.

nanoparticles in this case are polymer shells designed to produce a controlled release of a loaded anticancer drug and using aptamers coated on the surface to target the tumor cells. The nanoparticles were also labeled with fluorescent molecules and show up in the image as red dots. Clearly, the loaded nanoparticles have been taken up by the cells [4]. The same team showed that in rats with prostate cancer the nanoparticle treatment caused a complete eradication of the tumor in most cases [4].

A more versatile version of the lock-and-key targeting exhibited by aptamers is to use peptides, which are subunits of proteins and are composed of 20 amino acid building blocks. Thus they exhibit greater structural diversity due to the larger pool of building blocks and can be constructed to target a wider variety of antigens. As with all the other targeting molecules, the goal is to identify a suitable receptor or protein on the surface of a tumor cell that is specific to that kind of tumor or at least much more abundant on the tumor cell than on normal cells.

Finally, the smallest of the targeting molecules under investigation is folate (folic acid), consisting of just 51 atoms. This is a well-known vitamin (B9) and is useful as a targeting molecule because folate receptors are frequently overexpressed in a range of tumor cells. The main problem with this as a targeting

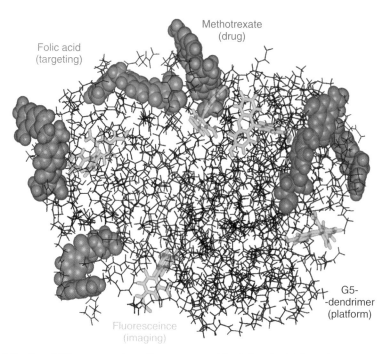

Methotrexate
(drug)

Folic acid
(targeting)

G5-
-dendrimer
(platform)

Fluoresceince
(imaging)

Fig. 6.8   Dendritic nanovector. Computer model of the dendritic carbon nanoparticle used as a carrier of an anticancer drug and targeted to tumor cells using folate molecules. Also attached are fluorescent molecules for diagnosis. These targeted drug carriers have shown a significant increase in anticancer activity and a decrease in the toxicity of the attached drug [5]. Image courtesy of Jolanta Kukowska-Latello, Michigan Nanotechnology Institute for Medicine and the Biological Sciences.

molecule is that folate receptors are also expressed in certain types of normal cells in the brain, placenta, lungs, intestines, and kidneys, so drug targeting using folate needs to be "fine-tuned" in order to optimize the selection of tumor cells.

A computer model of a nanovector loaded with folate targeting molecules and an anticancer drug is shown in Fig. 6.8. In this case the nanoparticle is a carbon dendrimer or spherical polymer and is acting as a passive carrier of both targeting molecules and drug so that the toxic anticancer agent is delivered only to tumor cells. For diagnostic purposes the nanoparticles can also carry fluorescent molecules, so their location relative to cell nuclei becomes apparent as in Fig. 6.7. Experiments on mice infected with a type of human skin cancer showed a significant increase in the antitumor activity of the drug and a decrease of its toxicity compared to the free drug [5].

6.1.6   Magnetic Targeting

A different form of targeting is to attach the drugs to magnetic nanoparticles that can then be concentrated at the correct site and kept there by an external magnetic field. A magnetic field gradient produces an attractive force on a magnetic particle, and the optimum size of the particle is ~100 nm—that is, the largest particle that is still a single magnetic domain (see Chapter 1, Section 1.2 and Advanced Reading Box 6.1). The particle, which is suspended within the circulation in the region of the tumor, has to be kept in position against the drag force of the moving fluid. It is simple to show (Advanced Reading Box 6.1) that a 100-nm-diameter iron particle would be washed away in even the strongest field gradient that can be applied from outside the body (~50 T/m) for a blood flow faster than about 15 μm/sec. This is slower than is found even in small blood vessels, so it would appear that this technique does not work; however, in practice the particles are pulled in one direction and quickly attach to some tissue in the region of the tumor and are concentrated there.

ADVANCED READING BOX 6.1—FORCE ON MAGNETIC NANOPAR-TICLES DUE TO A MAGNETIC FIELD GRADIENT

The force, **F**, on a magnetic particle with a total magnetic moment μ in a magnetic field **B** is given by

$$\mathbf{F} = (\mu.\nabla)\mathbf{B}$$

or, in one direction only, by

$$F = \mu \frac{\partial B}{\partial z}.$$

If the particle is a single domain particle, then μ is proportional to the particle volume; and so the larger the particle, the stronger the force. Above the critical size for the particle to remain a single domain, however, μ reduces to the average magnetization in a given direction of the different domains; and unless the particle is within a magnetic field strong enough to saturate it, its magnetization will be less than a single-domain particle in which the magnetization is permanently saturated. The optimum size is thus the largest particle that remains a single domain, that is, ~100 nm. If this were an Fe particle, it would contain about 4×10^7 atoms; and since the magnetic moment of a single Fe atom is 2.2 μ_B, the moment of the particle would be $\mu = 4 \times 10^7 \times 2.2 \times 9.274 \times 10^{-24} = 8 \times 10^{-16}$ A/m^2. In practice the largest field gradient that can easily be applied from outside the body is about 50 T/m so the maximum force that can be applied is 4×10^{-14} N. The drag force for a particle with a radius r in a fluid with a viscosity η moving with a velocity v is

$$F = 6\pi \eta v r.$$

The viscosity of blood is $\eta = 0.0027$ N $\cdot$ sec $/m^2$, so for a 50-nm radius Fe particle the magnetic force would be overcome by a blood flow faster than 15 μm/sec^{-1}. This is slower than in even small blood vessels, so in practice the technique relies on the particle being pulled to a tissue wall and sticking.

The main problem with the technique is that the magnetic field gradient, which is the parameter that determines the magnetic force on the nanoparticles, decays strongly with distance. With permanent magnets the figure of 50 T/m is only achievable to a depth of a few millimeters using a so-called Halbach array that concentrates the magnetic field on one side of the magnet assembly [6]. Recently, systems based on superconducting magnets that can produce usable field gradients at depths of tens of millimeters have been demonstrated [7].

The numbers above have assumed that the nanoparticles are pure iron, but these would be highly toxic and in general it is difficult to attach biological molecules to an iron surface. An alternative is iron oxide particles that are less toxic but also less magnetic. The ideal solution would be to have core-shell particles that consist of a pure iron core and a gold shell a few atomic layers thick as illustrated in Fig. 6.9. The shell would render the particle nontoxic, and there are well-developed methods to attach complex molecules to gold surfaces. On the other hand, the pure iron core ensures that the magnetic moment is as high as possible. Iron is the most magnetic element (at room temperature), but in more advanced designs of nanoparticle it might be possible to increase the magnetization of the core by making it out of more than one material (e.g., iron–cobalt). There have been several reports of core-shell particles produced both chemically and in the gas phase (see Chapter 4), so the functionalized magnetic carriers shown in Figure 6.9 are certainly feasible. In some cases the drug might not be fully effective while the nanoparticle is attached, but the magnetic carrier can be made to release the drug by applying a relatively weak external oscillating magnetic field, which heats the particle (Section 6.2.2).

A variant of the magnetic targeting in the blood is to use magnetic targeting to guide aerosol into the lungs. Many pulmonary diseases, including asthma, cystic fibrosis, and lung cancer, are treated by inhaled drugs; with conventional inhalers, only about 4% of the drug makes it through to the windpipe, so in order to get sufficient drug to where it is needed, some tissues have to be overexposed, leading to unwanted side effects. A recent study by a German consortium [8] tested aerosols that consist of tiny liquid droplets, with each one containing iron oxide nanoparticles with a diameter of 50 nm and a drug. They placed a magnetic probe next to one lung of a mouse and found that this lung received eight times more drug than its neighbor. Applying this to humans will be more difficult because of the larger lung volume and the difficulty already discussed above with getting a significant magnetic field gradient deep in the body. With

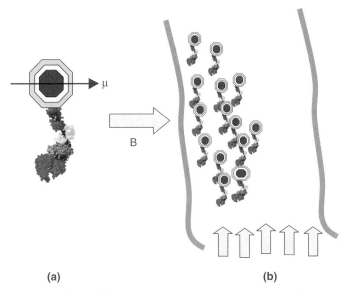

(a)   **(b)**

Fig. 6.9 **Magnetic drug delivery.** (a) Magnetic carrier nanoparticle, consisting of a magnetic core and coated in a shell that renders it non-toxic and easy to attach to biological molecules, loaded with a drug molecule. In more advanced designs the magnetic core can itself consist of more than one material to optimize its magnetic properties. (b) Loaded nanoparticle carriers in a blood vessel attracted to an external magnetic field source and trapped.

improved magnetic field generators [7] and magnetic particles with a higher magnetization than iron oxide, this may become a viable technology.

6.2 TREATMENT OF TUMORS BY HYPERTHERMIA

6.2.1 Biological Response to Heating

Hyperthermia is the term describing the elevation of the temperature of a living organism to a value high enough to result in changes or damage. A temperature increase of a few degrees from the normal human body temperature of 36–37°C to 42°C will start to induce malfunction of the biochemistry of cells, resulting in cell death.

Cell death is generally divided into two classifications—that is, apoptosis and necrosis. Apoptosis is a natural process in which the cells cooperate in their own death and is important in maintaining the health of a living organism. It is the method used to shut down and remove injured cells. In addition, apoptosis plays a vital role in developing fetuses to eliminate cells from unwanted regions—for example, to separate fingers and toes. A cell undergoing apoptosis in an orderly fashion shuts down and fragments in such a way that the debris is easily recognized by the innate immune system (see Section 6.1.4) and removed. In humans, anywhere between 2×10^{10} and 7×10^{10} cells are removed daily by apoptosis. Insufficient apoptosis can lead to cell proliferation and cancer, while an excessive amount can lead to atrophy—that is, the wasting away of part of the body.

In contrast, necrosis is the traumatic death of cells resulting from injury, infection, or poisons. It is less orderly and there is no signaling to nearby phagocytes to remove the dying cell, making it harder for the innate immune system to remove the debris. In addition, the cell membrane can rupture, thereby releasing the intracellular contents and resulting in swelling.

Gentle heating to $42–45°C$ stimulates apoptosis, though the precise mechanism is not well understood. One possibility is that newly synthesized proteins within the ribosomes (Fig. 6.4) are denatured—that is, not structured to fold correctly—and some of these are toxic to the cell. The important point, as far as tumor treatments are concerned, is that tumor cells are more sensitive to apoptosis in the temperature range $42–45°C$ than are normal cells. The observation that hyperthermia induced by high fevers can cure cancers was noticed even in ancient times, with the earliest known written record on the famous Edwin Smith Egyptian papyrus dating back to 3000 B.C. In 1893 Coley[9] carried out a controlled experiment in which a patient with tumors on his neck and tonsils was injected with a virulent strain of *Streptococcus* that induced erysipelas, a severe skin infection and fever. Within a few days the tumors had completely disappeared. Coley continued his research and tried to improve the fever-inducing toxins, eventually showing a 37% success rate in patients with inoperable tumours [9]. The nature of the treatment, however, invited ridicule and the method was not generally accepted by the medical community.

More recently, there has been a revival of interest in cancer treatment by hyperthermia stimulated by the development of methods where the heat from an external source is localized at the tumor. For example, focused microwaves [10] have been found to be an effective adjuvant to other, more established treatments such as radiotherapy. The possibility of targeting nanoparticles to tumors has increased interest further. If the particles are optically or magnetically active, they can be heated from outside the body by light of the correct wavelength or an oscillating magnetic field that does not harm normal tissue. This can be applied when the particles are in place on the tumor cells, guaranteeing that the hyperthermia is localized at the tumor. This new nanotechnology-based form of hyperthermia offers the promise of a side-effect-free gentle and generic cure for cancer, though this is some way off as a general treatment for humans.

The mechanisms that heat magnetic and optical nanoparticles by external electromagnetic fields are described in the following sections, but first it is worth examining in general terms how much power has to be absorbed by tissue to achieve the required temperature rise. Although we will assume that the nanoparticles are attached specifically to cancer cells by their targeting molecules, these

Table 6.1 Physical and Thermal Properties of Various Tissues and Blood

	Density ($kg \cdot m^{-3}$)	Specific heat ($J \cdot kg^{-1} \cdot K^{-1}$)	Thermal Conductivity ($W \cdot m^{-1} \cdot K^{-1}$)
Muscle	1026	4584	0.41–0.62
Fat	920–940	2230–2980	0.20–0.22
Blood (whole)	1060	3594	0.49
Liver		3411	0.47–0.56
Tumor		4090	0.51–0.56
Water (at 40°C)	992	4179	0.58

are intimately embedded in tissue or fluid and the heat generated will conduct efficiently into the environment. An elevated temperature of 42°C needs to be maintained in the tumor for about half an hour, and sufficient power needs to be generated to maintain the higher local temperature at the site against heat losses to the surroundings. The temperature of the region around the tumor will also increase, but ideally the temperature profile will drop rapidly below 40°C beyond the boundary of the tumor, minimizing apoptosis in the healthy tissue.

Figure 6.10a shows schematically the steady-state situation in which a tumor (the darker object) embedded in tissue is being heated by stimulating nanoparticles within it by an external electromagnetic field (to which biological tissue is transparent). Heat will be lost to the surroundings, whose temperature will increase, but ideally the combination of particle density and power dissipated can be optimized to give a temperature profile in which only the tumor cells are heated sufficiently to stimulate apoptosis. Data from an experiment on hyperthermia tumor treatment using magnetic nanoparticles in rabbits is shown in Fig. 6.10b. It is observed that after 30 minutes the tumor in the liver is maintained at a steady-state therapeutic temperature of about 43°C, while 1–2 cm away in healthy liver tissue the temperature is at about 38°C, which is too low to cause damage.

Within living tissue the heat flow is much more complicated than in a passive medium since the whole region is perfused with blood via microscopic blood vessels and the flow carries away heat. For any region larger than about a tenth of a millimeter, the blood perfusion can be considered as evenly distributed. In addition, there is a metabolic heat production within any small volume. The situation is complicated further by the fact that rises in temperature produce an increase in the blood perfusion rate (as the tissue attempts to maintain a constant temperature). The problem of heat flow in tissue with all these effects present was studied by Pennes in 1948 [11], who developed an equation to describe the temperature distribution, and this has been used ever since (see Advanced Reading Box 6.2). The most important tissue properties for calculating heats losses and temperature rises are listed in Table 6.1 for various tissues and blood. It is apparent that, with the exception of fat, most tissues have physical properties similar to those of water.

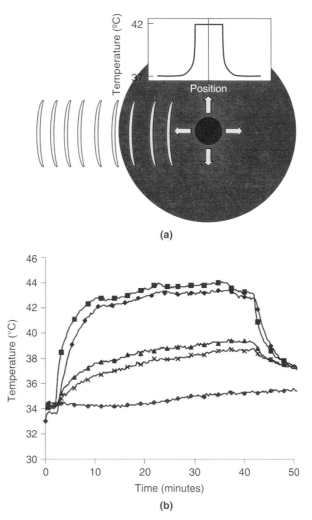

Fig. 6.10   Hyperthermia of a tumor embedded in healthy tissue. (a) Steady state situation in which a tumor (the darker object) embedded in tissue is being heated by stimulating nanoparticles within it by an external electromagnetic field (to which biological tissue is transparent). The ideal temperature profile in which only the tumor is at a temperature high enough to stimulate apoptosis is shown in the inset. (b) Data from hyperthermia treatment of liver tumors in rabbits using magnetic nanoparticles. The temperature has been measured at the center of the tumor (■), the edge of the tumor (♦) and two different sites in healthy liver tissue 1 – 2 cms away from the tumor (▲, X). The core body temperature is also shown (◇). Reproduced with the permission of Dr. Steve Jones at Sirtex.

ADVANCED READING BOX 6.2—HEAT DISSIPATION IN LIVING TISSUE

Heat dissipation within tissue that is being artificially heated in the presence of blood perfusion and metabolic heat generation is described by Pennes equation [11]:

$$\rho c \frac{\partial T(\mathbf{r}, t)}{\partial t} = \nabla.(k\nabla T(\mathbf{r}, t)) + c_b\omega_m(T_a - T(\mathbf{r}, t)) + Q_m + P(\mathbf{r}, t),$$

where:

ω_m is the blood perfusion rate in terms of mass (kg·m^{-3}·s^{-1})

ρ (kg·m^{-3}), c (J·kg^{-1}·K^{-1}), and k (W·m^{-1}·K^{-1}) are the density, specific heat, and thermal conductivity of the tissue, respectively

c_b is the density and specific heat of blood

$T(\mathbf{r},t)$ is the local temperature and T_a is a reference temperature (e.g., the arterial blood temperature)

Q_m is the metabolic heat production rate per unit volume (in W·s^{-1})

$P(\mathbf{r},t)$ is the spatially distributed artificial heat deposited per unit volume (in W·s^{-1})

Every term on both sides of Pennes equation has units W·s^{-1}. If we assume that the metabolic rate, Q_m, is constant (in fact it is temperature-dependent as is the blood perfusion rate, but we will ignore this in the simplified analysis here), we can subtract the steady-state temperature field and rewrite the Pennes equation in terms of $\theta(r,t) = T(\mathbf{r}, t) - T_a$, that is, the difference between the local temperature and the reference temperature (core body temperature). The result is

$$\rho c \frac{\partial \theta(\mathbf{r}, t)}{\partial t} = \nabla.(k\nabla\theta(\mathbf{r}, t)) - c_b\omega_m\theta(\mathbf{r}, t) + P(\mathbf{r}, t).$$

Let us now assume that the externally stimulated power generation has been applied for a sufficiently long time that the temperature is no longer changing ($\partial\theta/\partial t = 0$). We will also assume that the properties of the tissue are isotropic so that we can consider the temperature change in one direction only. Thus away from the boundary where P goes to zero, the temperature variation is given by

$$\frac{\partial^2\theta(x)}{\partial x^2} = \frac{c_b\omega_m}{k}\theta(x).$$

Writing $\lambda^2 = k/c_b\omega_m$, it is apparent that the temperature at the heated boundary decays exponentially according to

$$\theta(x) = \theta_0 e^{-x/\lambda}$$

with a characteristic decay length λ The temperature θ_0 is the temperature at the edge of the heated region. In order to find this, we need to solve the full equation and determine the asymptotic value of θ. A rule of thumb is that an applied power density of $P = 0.1$ W·m^{-3} will produce the required temperature rise of 5 K relative to the reference temperature ($\theta = 5$ K). Typical values for k, c_b, and w_m are $k = 0.5$ W·m^{-1}·K^{-1}, $c_b = 4000$ J·kg^{-1}·K^{-1}, and $w_m = 5$ kg·m^{-3}s^{-1}, so the characteristic decay length of the temperature from its heated value to zero (i.e., return to the reference temperature) is $\sqrt{0.5/(4000\times5)}$ m or about 5 mm, which is in reasonable agreement with the data shown in Fig. 6.10b.

As a rule of thumb, the power required to heat tissues to a therapeutic temperature is ~ 0.1 W/cm^3. Thus active nanoparticles have to be dispersed within the tumor at a sufficient density to achieve this heating power when stimulated by externally applied electromagnetic fields. The most useful measure of the effectiveness of nanoparticles in producing heat is their specific absorption rate (SAR), which is the power (in watts) generated by 1 g of nanoparticles in response to a particular stimulation. This can be an oscillating magnetic field in the case of magnetic nanoparticles or an infrared beam for Au nanoshells (Section 6.2.3) or carbon nanotubes (Section 6.2.4). In all cases the SAR is sensitive to the frequency of the applied stimulation. To give some idea of the nanoparticle density required, with currently available magnetic nanoparticles based on Fe oxide, an SAR of 10 W/g is easily attainable. Combined with the rule of thumb that about 0.1 W per cubic centimeter is required to heat and maintain tissue at the therapeutic threshold of 42°C, it is evident that 0.01 g (10 mg) of nanoparticles have to be dispersed per cubic centimeter of tissue. This is a low density in general terms; but if targeted nanoparticles are being used, another important parameter is the number of nanoparticles per tumor cell since there are a finite number of receptors on each cell. If the nanoparticles used were 10-nm-radius Fe oxide particles, then there would be 3×10^{14} nanoparticles in 10 mg. The density of tumor cells in a cancer is of the order of 10^8 per cubic centimeter; thus hundreds of thousands of nanoparticles with an SAR of 10 would be needed per tumor cell to achieve the required heating power. Although some types of receptor are expressed in these numbers on the cell surface, this may not be the case for the specific receptor chosen for targeting, which could be of the order of a thousand. So for targeted nanoparticle hyperthermia to work, it is clear that an SAR performance of much higher than 10 is required. As discussed in the following sections, large improvements in SAR performance are expected with the availability of new types of magnetic nanoparticle.

6.2.2 Magnetic Nanoparticle Hyperthermia

The basic idea of using magnetic particles to heat tissue dates back over 50 years [12]. Recently, enormous improvements in the technique have been achieved with the production of nanoparticles with good size control coupled with their attachment to targeting moieties so that the heat is localized at a cellular level within the cancer.

There are two types of magnetic heating mechanism for nanoparticles in an alternating magnetic field, and which one applies depends on their size. Any particle that would be used for hyperthermia will be smaller than 100 nm and thus would be a single magnetic domain. This means that all the atomic magnetic moments are locked together to form a single giant magnetic moment, and the whole particle can be thought of as a permanently magnetized bar magnet whose magnetization can be reversed by applying a magnetic field. It is important here to distinguish between (a) the particle itself rotating, which it can do if it is floating freely in a fluid and (b) the particle immobilized, and the magnetization rotating within it, which would occur if it were chemically attached to a cell (the normal situation in hyperthermia). Assuming the latter situation, there is an energy barrier, the anisotropy barrier, separating the two different magnetisation directions, which has to be overcome to cause the magnetization to flip. The size of this energy barrier produces two distinct types of behavior in magnetic nanoparticles. If the barrier is very large relative to the average atomic thermal energy in the system at blood temperature, then the particle magnetization is stuck or "blocked" in a given direction until it is reversed by an external field. If, on the other hand, the energy barrier is very small compared to the atomic thermal energy, the magnetization will constantly be driven to either direction at very high frequency by the innate thermal energy within the particle. In this state, its average magnetization will be zero until an external magnetic field is applied and then the particle magnetization will be biased in one direction by the field. In this state the particles are *superparamagnetic*. The size that separates the blocked state from the superparamagnetic one depends on the material and on the temperature; but, for example, for a spherical Fe nanoparticle at blood temperature it is at a diameter of about 7 nm. Figure 6.11 illustrates schematically the difference between the two states.

We will start by considering the heating of blocked particles. Figure 6.12 shows one of the highly functionalized magnetic nanoparticles, which is large enough to be blocked, attached to its target cell where it is immobilized. Assume an alternating magnetic field is applied from outside the body that is large enough to flip the magnetization within the particle at the same frequency. Every time the particle magnetization reverses, an amount of energy equal to the anisotropy barrier is released into the surroundings. A rough estimate of the heating power supplied by the particle is given by the product of the anisotropy barrier times the frequency of the magnetic field. For example, if the central magnetic core were a 20-nm-diameter cobalt nanoparticle, the anisotropy barrier would be about 2×10^{-18} J, and thus driving the magnetization at 100 kHz would produce a

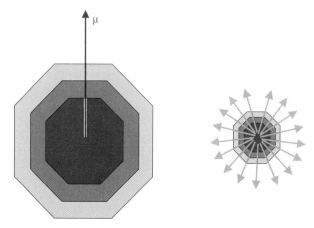

Fig. 6.11 Superparamagnetic and blocked magnetic nanoparticles. Illustration of a magnetic nanoparticle that is large enough (>10 nm diameter) to be magnetically blocked at blood temperature so it is permanently magnetized in one direction compared to a smaller superparamagnetic nanoparticle in which thermal energy flips the moment around randomly at high frequency.

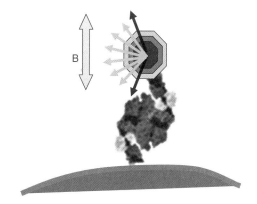

Fig. 6.12 Heating blocked magnetic nanoparticles by an applied AC magnetic field. A blocked magnetic nanoparticle with attached targeting molecule latched onto the membrane of a target cell. An alternating magnetic field forces the magnetization to switch back and forth producing heat in the nanoparticle (see Advanced Reading Box 6.3).

power of about 2×10^{-13} W (0.2 picowatts). This is a tiny amount of power, but cells are small objects and a 10-μm-diameter cell would only need about 1000 of these particles with these conditions to heat up by 10°C in 5 minutes. One thousand 20-nm-diameter nanoparticles would constitute about 0.0004% of the cell volume. Incidentally, cobalt nanoparticles have been chosen for this example because they have a higher anisotropy barrier than do iron nanoparticles of the same size. Advanced Reading Box 6.3 presents a more rigorous calculation of the heating power derived from blocked particles.

To compare performances of various types of nanoparticles we need to determine their SAR. Using the above figures as an example, a 20-nm-diameter cobalt nanoparticle weighs 3.73×10^{-17} g and with each one producing 2×10^{-13} W the SAR figure is thus $2 \times 10^{-13}/3.73 \times 10^{-17} = 5600$ W/g. This is a pretty impressive figure but unattainable in practice because the frequency and amplitude of the magnetic field required would be large enough to have a harmful effect on normal tissue (see Advanced Reading Box 6.3). In this particular example a large-amplitude field is required to magnetize the particles; and in order to keep the amplitude–frequency product at a safe value, we would have to substantially reduce the frequency and thus the power generated per particle. As shown in Advanced Reading Box 6.3, the SAR we could derive in practice from the blocked Co nanoparticles would be more like $\sim$30 W/g. Alternatively, we could keep the high frequency and lower the field amplitude, but then the particles would only reach a fraction of their saturated magnetization at each cycle, again reducing the SAR. For hyperthermia it is better to use smaller superparamagnetic nanoparticles, which generate heat via a different mechanism outlined below.

ADVANCED READING BOX 6.3—HEATING POWER FROM BLOCKED NANOPARTICLES

Permanently blocked particles display an open hysteresis loop as in the figure, which is obtained from a dilute assembly of 10-nm Co nanoparticles.

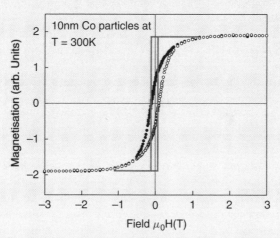

Each cycle around the hysteresis loop generates an amount of energy proportional to the area in the loop; that is,

$$\mu_0 \oint H dM$$

per unit volume, so at frequency f the power generated per unit volume is

$$P_{Bl} = f\mu_0 \oint HdM.$$

The problem in general is that the detailed shape of the hysteresis loop depends on the fine details of the particles, including their precise shape. To get an estimate of the power derived from the particles whose magnetization is shown in the figure, one can draw an equivalent square loop shown in red. The amplitude of this loop is then the volume (per cubic meter) saturation magnetization of Co $(1,352,642$ A m$^{-1})$ and its width is twice the coercive field, which from the figure is ~ 0.2 T. The energy per cycle per unit volume is thus 270,000 J; and if we drive the field at 100 kHz, then from the density of Co (8910 kg m^{-3}) the SAR of the particles would be ~ 3000 W/g, which is of the same order as the rough estimate in the text. In practice, it is generally recognized that a field-frequency $(H \times f)$ product of about 4.85×10^8 A m^{-1}s^{-1} is the maximum safe dose for extended periods to avoid damage to normal tissue. In the hysteresis curve shown above, a field of $\sim 400,000$ A m^{-1} $(= 0.5T/\mu_0)$ is required to approach saturation; this means that the maximum safe frequency we could apply would be ~ 1 kHz, which brings the SAR down to ~ 30.

With blocked particles the heating due to driving the magnetization in an immobile particle is only one mechanism. The particle will not be totally immobile and will bend its attached molecule as its magnetization is reversed. This shaking of the system provides heat. In addition, particles floating in suspension around the target cells and not yet attached will be shaken, thus generating additional heat in the fluid between the tumor cells.

For blocked particles the fundamental heating mechanisms are (a) the driving of the particles magnetization vector backwards and forwards across its internal energy barrier and (b) the agitation of the particle within the fluid. The energy barrier in small superparamagnetic nanoparticles is insignificant; but, when driven at a suitable frequency, these nanoparticles are capable of generating impressive SARs. In addition, whereas blocked particles require relatively large field amplitudes (and hence low frequencies to minimize direct tissue heating), superparamagnetic particles can generate heat at small applied field amplitudes so that high frequencies can be used. The heating mechanism for small particles is a subtle thermodynamic process and relies on the fact that in a high-frequency magnetic field the particles magnetization is always trying to catch up with the applied field.

The basis of the heating mechanism can be understood with reference to Fig. 6.13. Consider an assembly of superparamagnetic nanoparticles that is placed within a sufficiently strong magnetic field to force all the normally fluctuating moments to point in the same direction so that the sample is magnetically saturated. If the field is suddenly removed, the system will go from a highly ordered

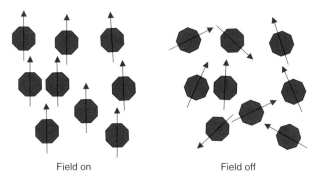

Field on   Field off

Fig. 6.13 Heating mechanism for superparamgentic nanoparticles. If a group of superparamagnetic nanoparticles is magnetized by a strong external field (a) and the field is suddenly removed (b) the particles will quickly go from a magnetically ordered to a disordered state and in the process remove heat from the heat bath in which they are submerged. This is the heat energy required to start their magnetic moments randomly fluctuating. Going from (a) to (b) will thus produce cooling (magnetic refrigeration). Clearly going the other way, that is, suddenly applying a magnetic field to drive the assembly from the disordered to the ordered state will heat the system. Driving the nanoparticles with a high frequency oscillating magnetic field will continuously pump heat into the system.

(low entropy) state to a disordered (high entropy) state; in the process, heat will be removed from the heat bath in which the particles are immersed. One can see this more simply as heat removed from the system to get the magnetic moments randomly oscillating. In fact this process is used for magnetic refrigeration. Conversely, if the assembly is suddenly magnetized from the disordered state, heat will be added to the system. As we drive the system in a high-frequency oscillating magnetic field, the fact that the magnetization lags the applied field means that heat is continuously pumped into the system. Advanced Reading Box 6.4 derives an equation for the power generated by driving superparamagnetic nanoparticles with a high-frequency applied magnetic field. Suspensions of superparamagnetic nanoparticles are referred to as ferrofluids.

ADVANCED READING BOX 6.4—HEATING POWER SUPERPARAM-AGNETIC NANOPARTICLES

To calculate the power dissipation, we start with the first law of thermodynamics:

$$dU = \delta W + \delta Q,$$

where U is the internal energy, Q is the heat added, and W is the work done. For an adiabatic process, $dQ = 0$ and the magnetic work is given by $dW = \mathbf{H}.d\mathbf{B}$, where $\mathbf{H}$ is the magnetic field intensity (in A/m) and $\mathbf{B}$ is the induction (in T).

so the above equation becomes

$$dU = \mathbf{H}.d\mathbf{B}.$$

$\mathbf{H}$ and $\mathbf{B}$ are always in the same direction so that only magnitudes are required and $B = \mu_0(H + M)$, so the change in internal energy for each cycle is given by

$$\Delta U = \mu_0 \oint H d(H + M)$$

Integrating by parts and bearing in mind that H integrated over a cycle is zero yields

$$\Delta U = -\mu_0 \oint M dH$$

When the magnetization lags the field, the integrand is negative, so the above equation yields a positive result showing that magnetic work is converted to internal energy. When dealing with ferrofluids, it is convenient to express the magnetization response as a complex susceptibility defined by

$$\chi = \chi' - i\chi''$$

If the applied magnetic field is sinusoidal and expressed as

$$H(t) = H_0 \cos \omega t = \mathrm{Re}[H_0 e^{i\omega t}],$$

Then since $M = \chi H$, the magnetization is as follows from the previous two equations:

$$M(t) = \mathrm{Re}[\chi H_0 e^{i\omega t}] = \mathrm{Re}[(\chi' - i\chi'')(\cos \omega t + i \sin \omega t)]$$
$$= H_0(\chi' \cos \omega t + \chi'' \sin \omega t),$$

which makes it clear that χ' is the in-phase component while χ'' is the out-of-phase component. From the field dependence we have $dH = -H_0 \omega \sin \omega t$, so changing the variable to t in the equation for ΔU and substituting M from the previous equation gives

$$\Delta U = \mu_0 \omega H_0^2 \chi'' \int_0^{2\pi/\omega} \sin^2 \omega t \, dt$$

(since the integral over 1 cycle of $\cos \omega t \sin \omega t$ is zero). The definite integral has value π/ω and multiplying the internal energy per cycle by the frequency yields the power P generated by the assembly of nanoparticles in response to the applied oscillating magnetic field, given by

$$P = \mu_0 \pi f H_0^2 \chi''.$$

Table 6.2 shows the SAR performance of some available ferrofluids after scaling to "safe" strength–frequency combinations of the applied field (see Advanced Reading Box 6.3). For superparamagnetic nanoparticles it is possible to derive an exact expression for the power absorbed as a function of the particle size [13]. Figure 6.14 shows the SAR, calculated for pure Fe nanoparticles as a function of the particle diameter. The curves are shown for three different applied field frequencies, where, for each frequency, the field amplitude has been set to the maximum safe value. It is clear that it is possible to obtain much higher SAR figures for certain particle sizes and applied field conditions than for the available oxide nanoparticles. Pure Fe nanoparticles cannot be used, however, because they are highly toxic and require a biocompatible coating for medical applications. A significant effort is devoted to researching methods to produce core-shell particles with a highly magnetic core, which gives a good SAR performance and a biocompatible inert shell onto which it is easy to attach targeting moieties.

Up to now, this technology has only been tested without biological targeting in which the magnetic nanoparticles are injected into the region of the tumor and the external oscillating field is applied before the particles disperse, but it

Table 6.2   SAR Figures for Various Ferrofluids [14]

Nanoparticle Material	Diameter (nm)	Particle Coating	Applied Magnetic Field[a] Amplitude[a] (kA/m)	Applied Magnetic Field Frequency (kHz)	Dispersion Medium	SAR (W/g)	Reference
Fe_3O_4	8	None	1600	300	Water	5	15
Fe_3O_4 (Endorem®)	6	Dextran	1600	300	Water	<0.1	15
Fe_2O_3	3	Dextran	970	500	Water	8	16
Fe_2O_3	5	Dextran	970	500	Water	41	16
Fe_2O_3	7	Dextran	970	500	Water	49	16
Fe_2O_3[b]	10–12	Dextran	557	880	Water	16	17
Fe_2O_3[b] (MFL 082AS®)			3000	100	Water	~100	
Pure Fe[c] (calc. see Fig. 6.14)	13.4	None	4853	100	Water	138	

[a] The field amplitudes have been scaled down from those used in the published experiments so that the amplitude × frequency product is at the maximum "safe" value (see Advanced Reading Box 6.3).
[b] These particles have been used in tests of hyperthermia in animals or humans.
[c] Pure Fe nanoparticles would be highly toxic, so they would have to be coated with a biocompatible layer before use.

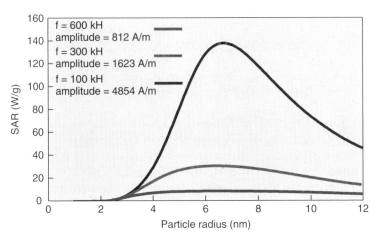

Fig. 6.14 SAR vs. size for Fe nanoparticles. Calculated SAR [13] *vs.* particle radius for pure Fe nanoparticles in water under different applied field conditions. Note that for each frequency the field amplitude has been set at the maximum safe level (see Advanced Reading Box 6.3). Clearly these can have a very high performance compared to Fe oxide nanoparticles (Table 6.2), however pure Fe particles are highly toxic and would need a biocompatible coating before use.

clearly shows great potential. Figure 6.15 shows the results of a study of magnetic nanoparticle hyperthermia in mice [18], each bearing a sarcoma (cancer of connective tissue). The data are from four different groups of animals, including a control group (group 1) that received no treatment, a group that received magnetic nanoparticle hyperthermia but the nanoparticles were allowed to disperse naturally (group 2), and two groups in which the nanoparticles were held in place by an external constant magnetic field (groups 3 and 4). The tumor growth in group 2 was slowed relative to the control group, but in groups 3 and 4 complete regression was found in 33% of the mice. The nanoparticles used in this study were dextran-coated Fe_2O_3 particles with a magnetic core diameter of 10–12 nm and an SAR of about 16 (scaled to the "safe" applied field value). The actual field–frequency product used in this study, however, was about 17 times the "safe" value and far too high to be applied to human patients.

The method is developing fast, and there now exists a commercial system for hyperthermia treatments (see Fig. 6.16) with phase II clinical trials on human patients underway. During phase I trials, designed to check for side effects and problems related to the therapy itself, recurrent tumors in the brain and prostate that did not respond to other treatments and for which the prognosis was poor were treated [19, 20]. These studies have produced important information related to the therapy. For example, the field–frequency product applied to the prostate region has to be limited to the "safe" value (see Advanced Reading Box 6.3) to avoid causing discomfort in patients. On the other hand, much higher fields—up

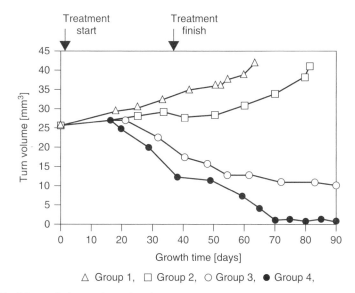

Fig. 6.15 Magnetic hyperthermia of mouse sarcoma. Results from a trial of magnetic nanoparticle hyperthermia on a sarcoma in mice. The data shows how the tumor volume develops with time for four different groups of animals. Group 1 received no treatment, group 2 were given treatment but the particles were allowed to disperse naturally. In groups 3 and 4 the nanoparticles were held in place by an external constant magnetic field. These groups experienced a complete regression of the tumor in 33% of cases. Reproduced with the permission of Elsevier Science from N. A. Brusentsov et al. [18].

to twice the "safe" value—can be applied to the brain region without any harmful effects. It is thought that the reason for this is that the prostate tumor lies within a complex morphology of bone and soft tissue and the sharp changes in dielectric constant produce reflections of the applied electromagnetic field that produce local "hot spots." In contrast, the brain tumor lies within a homogeneous mass of soft tissue that is transparent to the electromagnetic field. In all cases the patients tolerated the treatment well; and although the phase I studies were conducted on small groups and not designed to determine the efficacy of the therapy, there were some promising outcomes for the therapy. For example, in the case of the brain tumors, which were of a very aggressive type, a combination of hyperthermia combined with radiotherapy produced longer survival rates than would be expected without the hyperthermia, and one of a group of 14 patients experienced a complete remission. One of the problems encountered by the phase I studies is that it is difficult to achieve the required temperature rise throughout the whole volume of the tumor without using unacceptably high applied fields. With improvements in the SAR values of ferrofluids, however (Table 6.2 and Fig. 6.14) this method has the potential to be a very effective treatment for cancer. An important aspect of magnetic nanoparticle hyperthermia is that the same nanoparticles will act as contrast enhancers in magnetic

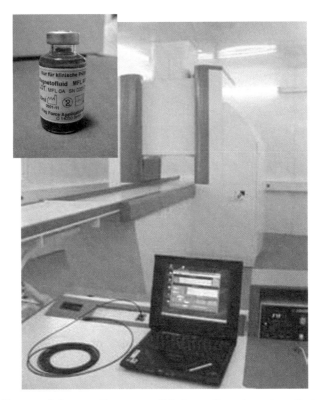

Fig. 6.16 **Commercial magnetic nanoparticle hyperthermia system.** System marketed by MagForce Nanotechnologies AG in Berlin. The patient lies on the moving bed and the electromagnet can apply a 100kHz alternating magnetic field with field strengths up to over 10,000 Am^{-1}. (about twice the 'safe' value) to any part of the body. The inset shows a bottle of the commercially available ferrofluid MFL 082AS® (see Table 6.2) with a very high SAR of about 100 Wg^{-1} at the 'safe' applied field. Reproduced with permission from MagForce Nanotechnologies AG.

resonance imaging (MRI) (Section 6.3.1), enabling a sensitive determination of where the particles are within the body. In the language of this chapter they are simultaneously "magic beacons" and "magic bullets."

6.2.3 Hyperthermia with Au Nanoparticles

The basic idea of "switching on" the hyperthermia by applying an external influence when nanoparticles are in place also applies to other types of energy absorption. Essentially the method relies on using electromagnetic waves that are heavily absorbed by the nanoparticles but for which tissue is transparent. Magnetic hyperthermia discussed in the last section, for example, applies electromagnetic waves with frequencies in the range 100 kHz to 1 MHz whose

Fig. 6.17 Surface plasmon resonance (SPR) in Au nanoparticle. Illustration of a Surface Plasmon Resonance (SPR) in a Au nanoparticle. The electrons of the particle, represented by the orange cloud, undergo a collective oscillation about the Au atom cores, represented by the green spheres. Reprinted from the University of Rostock Cluster Group homepage (http://www.physik.uni-rostock.de/cluster/) with permission from Prof. Karl-Heinz Meiwes-Broer.

wavelength is long compared to human dimensions and are not absorbed unless magnetic nanoparticles are present. Living tissue is also relatively transparent to near-infrared light [25], and so nanoparticles that are strong infrared absorbers can be used for hyperthermia.

Au nanoparticles are very strong absorbers of visible light due to a resonant condition in which the electric field of the light causes the whole electron cloud of the nanoparticle to oscillate coherently relative to the atomic background as indicated in Fig. 6.17. This condition is known as the surface plasmon resonance (SPR).

In general the interaction between the particle and an incident beam of light can be characterized in terms of its extinction—that is, the light it removes from the incident beam. As illustrated in Fig. 6.18, there are two processes that remove light from the incident beam, namely, absorption and scattering. For example, the beam can excite the SPR shown in Fig. 6.17 and the SPR itself can generate heat, in which case the removed light has gone into absorption. It also acts as a dipole shaking at the frequency of the incident light and is itself a source of light, and in this case the removed light is converted to scattered light. Both these processes are useful in medical applications since a single particle can scatter enough light to be visible in an optical microscope and report its position, and the heat generated by absorption can be used for hyperthermia.

The extinction by the particle depends not only on the optical properties of its constituent material but also on those of the medium in which it is embedded. For example, Fig. 6.19 shows the extinction efficiency for a gold nanoparticle of radius 10 nm immersed in air and in blood. As discussed in Advanced Reading Box 6.5, an extinction efficiency of 1 represents a completely opaque disk with the same size as the particle blocking the beam, but for the strong extinction by the SPR the extinction efficiency can become higher than 1 (see Fig. 6.20) and the particle can behave as if it were a larger opaque object in the beam.

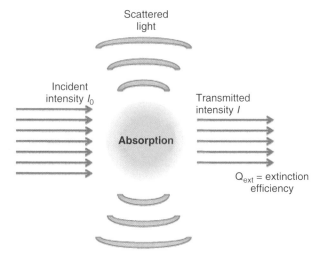

Fig. 6.18 Light extinction by a nanoparticle. After encountering a nanoparticle, incident light with an intensity I_0 is reduced to an intensity I and the amount of light extinguished is given by the extinction efficiency Q_{ext}, (see Advanced Reading Box 6.5). The lost light is either absorbed by the particle or converted to scattered light. The absorbed light is converted to heat and can be used for hyperthermia while the scattered light can be used to detect the location of the particle.

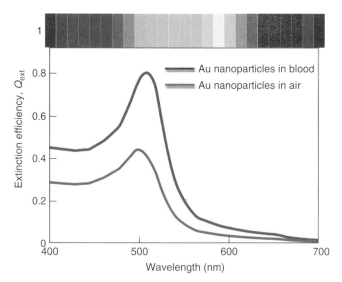

Fig. 6.19 Extinction by Au nanoparticles. Extinction efficiency of Au nanoparticles with a radius of 10 nm in blood and in air as a function of the wavelength of the incident light. In this case the strong extinction peak produced by the SPR is in the visible part of the spectrum and the bar across the top shows the color corresponding to the wavelength plotted on the x-axis.

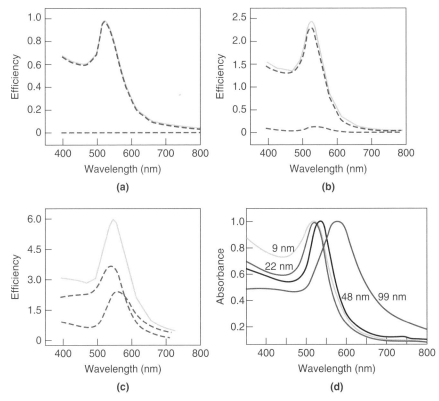

Fig. 6.20   Extinction vs. size for Au nanoparticles. Calculations of the absorption (red dashed curves), scattering (blue dashed curves) and extinction efficiency, Q_{ext}, (green solid curves) as a function of the wavelength of the incident light for Au nanospheres of diameter (a) 20 nm, (b) 40 nm, and (c) 80 nm. (d) Measured extinction spectra of Au nanoparticle suspensions of different diameters varying from 9-99 nm. Note that up to a diameter of ~50 nm the position of the peak is independent of particle size (see Advanced Reading Box 6.5). (Parts (a)-(c) reprinted with permission from [23] © 2006 American Chemical Society. Part (d) reprinted with permission from [24] © 1999 American Chemical Society).

ADVANCED READING BOX 6.5—EXTINCTION BY THE SPR IN GOLD NANOPARTICLES

The absorption and scattering of light by small particles is described by Mie theory [21] and involves solving the electromagnetic wave equations in the presence of a spherical boundary. The full theory is beyond the scope of this book; but for particles whose radius, R, is much smaller than the wavelength of the incident light, λ, the extinction cross section is given by

$$C_{ext} = \frac{24\pi^2 R^3 \varepsilon_m^{3/2}}{\lambda} \frac{\varepsilon_2}{(\varepsilon_1 + 2\varepsilon_m)^2 + \varepsilon_2^2},$$

where ε_1 and ε_2 are the real and imaginary parts of the complex dielectric function $\varepsilon = \varepsilon_1 + i\varepsilon_2$ of the constituent material of the nanoparticles and ε_m is the dielectric constant of the of the surrounding medium, which is assumed to be a real number (for air $\varepsilon_m = 1$ and for blood $\varepsilon_m \approx 1.4$). The dielectric function for the particle material in the required wavelength region can be obtained from published tables [22]. The equation predicts a maximum in the extinction at a wavelength for which the condition $\varepsilon_1 = -2\varepsilon_m$ is satisfied. Note that this depends only on the material and not the size of the particle, so the wavelength of the strong SPR peak according to the equation does not shift with particle size; however, for particle sizes such that $R > 0.1\lambda$ ($\sim$50 nm for visible light) the equation starts to become invalid and the position of the SPR is size-dependent for larger nanoparticles (see Fig. 6.20). The extinction efficiency Q_{ext} defined in the text is C_{ext} divided by the geometrical cross section of the particle (πR^2) and is given by

$$Q_{ext} = \frac{24\pi R \varepsilon_m^{3/2}}{\lambda} \frac{\varepsilon_2}{(\varepsilon_1 + 2\varepsilon_m)^2 + \varepsilon_2^2}.$$

One would expect the maximum value of Q_{ext} to be 1 corresponding to an opaque disk with the same radius as the particle, but the particle can diffract light away from the original direction of propagation as well as block it, so a detector along the line of incidence can register an extinction efficiency >1.

The relative amount of extinction and absorption depends on the particle size as shown in Fig. 6.20. In small particles (20 nm) the extinction is almost entirely due to absorption, while in the largest particles (80 nm) absorption and scattering both contribute approximately equally. Thus small particles are good for hyperthermia, and large particles are good for diagnosis and there is an optimum size to perform both functions.

Although the relative proportion of scattered and absorbed light can be varied as a function of particle size, it is evident from Fig. 6.20 that the wavelength of the resonance for pure Au spherical particles remains in the visible region of the spectrum. At these wavelengths, skin is highly absorbing and the penetration depth is only a few millimeters. The wavelength of the resonance needs to be shifted to the near-IR region, at which the penetration depth of tissue is several centimeters [25].

This can be achieved by varying the shape of the particles—for example, producing rod-like morphologies (nanorods)—or by using core-shell particles

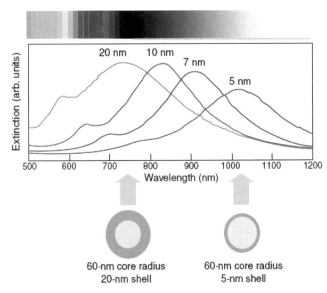

Fig. 6.21 Extinction by core-shell nanoparticles. Extinction vs. wavelength for nanoparticles consisting of a 60 nm diameter silica core and different thicknesses of Au shell. Varying the shell thickness from 20 nm to 5 nm shifts the wavelength of the resonance from the visible to the infrared region of the spectrum. Reprinted with permission from [26] © 2004 Adenine Press.

consisting of a dielectric core coated in a gold shell (nanoshells). For different morphologies or internal particle structures, there is still a strong resonance, but the wavelength at which it occurs can be tuned over a wide range. For example, Fig. 6.21 shows the extinction as a function of wavelength for nanoparticles consisting of a silica core and different thicknesses of gold shell. Varying the shell thickness from 20 nm to 5 nm shifts the wavelength of the resonance from the visible to the infrared region of the spectrum [26].

The SPR can be tuned over a similar range by changing the aspect ratio (length/radius) of Au nanorods as shown in Fig. 6.22a [27]. As with the core-shell particles, control over the geometry of the nanorods allows the wavelength of the SPR to be shifted from the visible to the near-infrared region. The shift in the wavelength of the SPR is easily seen by the changing color of suspensions of Au nanorods with different aspect ratio as displayed in the inset in Fig. 6.22b [28].

A number of *in vitro* studies of cancer cell death following exposure to infrared light after attaching Au nanoshells or nanorods have been conducted [29]. A commonly used light source is a titanium:sapphire laser, which emits at a wavelength of 800 nm; and as can be seen in Figs. 6.21 and 6.22, this matches the wavelength of the SPR in Au nanoshells with a core diameter of 60 nm and a shell thickness of 10 nm or Au nanorods with an aspect ratio of 3.9. The power required to cause damage in cancer cells with appropriate nanorods or nanoshells attached is much less than for healthy unlabeled cells [29].

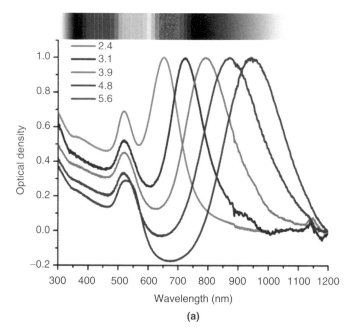

(a)

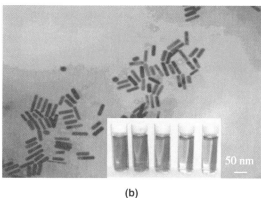

(b)

Fig. 6.22   Extinction by Au nanorods. (a) The wavelength of the SPR in Au nanorods as a function of the aspect ratio (length/radius) of the rods. (b) TEM image of Au nanorods with an aspect ratio of 3.9 (inset) colloidal suspensions of Au nanorods of increasing aspect ratio from left to right [28]. Reprinted with permission from [27] © 2006 American Chemical Society. Inset reproduced with the permission of Elsevier Science from L. M. Liz-Marzan [28].

6.2.4   Hyperthermia with Carbon Nanotubes

Specific types of single-walled carbon nanotubes (SWNTs; see Chapter 3, Section 3.9) are also powerful absorbers of near-infrared light, and they exhibit special properties that have stimulated great interest in their application to medicine.

They have been shown to be very effective at transporting molecular cargoes across the cell membrane into the cell interior and are nontoxic. With targeting, either they can be used to deposit required drugs, proteins, and so on, into specific cells or their high infrared absorbance can be used to kill the targeted cells by hyperthermia. Their utilization is illustrated in Fig. 6.23, which demonstrates the effectiveness of SWNTs at killing targeted cells [30]. This study used HeLa cells that originated from a cervical cancer biopsy in 1951 and have been propagated since for medical research. They have several beneficial attributes as model systems for research, including their rapid multiplication and "immortality"; that is, they can keep dividing an unlimited number of times. By culturing them in suitable conditions, it is also possible to stimulate them to overexpress folic acid receptors on their surfaces (denoted FR$^+$ cells) and thus one can do comparative studies with identical cells apart from the presence of folic acid receptors.

The SWNTs with average length 150 nm were conjugated with a phospholipid and a PEG molecule (see Section 6.1.2) onto the end of which was attached either folic acid (Section 6.1.5 and Fig. 6.8) or a fluorescent marker to reveal the position of the nanovector. The folic acid terminated nanovectors target the FR$^+$ cells and are transported through the cell membrane by endocytosis (Section 6.1.3) as illustrated schematically in Fig. 6.23b. The success of the targeting is revealed by comparing the confocal microscope images of SWNTs attached to both folic acid and fluorescent marker end groups. In the case of the FR$^+$ cells, large numbers of the nanovectors have entered the cells and are clustered around the cell nucleus (Fig. 6.23d), whereas there are few nanovectors internalized in the normal HeLa cells (Fig. 6.23e). FR$^+$ and normal HeLa cells were incubated with folic acid terminated nanovectors and then exposed to infrared light with a wavelength of 808 nm and an intensity of 1.4 W/m^2 for 2 minutes. It is evident that the FR$^+$ cells are killed by this treatment, whereas the normal cells are still functioning as healthy cells.

The same study also demonstrated that carbon nanotubes, once inside the cell, could be stimulated by exposing to infrared light to "dump" a cargo of DNA that could then cross the nuclear membrane into the nucleus. This leads to the potential of genetically engineering selected cells.

Thus carbon nanotubes have great potential as the basis for a potent technology for cancer treatment, but as infrared absorbers they suffer from the same fundamental disadvantage of the Au nanoparticles; that is, tissue is not sufficiently transparent for the treatment to be effective at depths greater than about 4 cm below the skin. A recent study [31], however, shows that nanotubes are also strong absorbers of radio waves at 13.56 MHz, at which tissue is highly transparent. The absorption mechanism is not clear, but the method is under intense scrutiny and nontargeted nanotubes have already been tested in rabbits bearing hepatic tumors with very encouraging results. This could thus be an alternative to the more highly developed magnetic nanoparticle hyperthermia (see Section 6.2.2) but with the additional advantages offered by carbon nanotubes such as their ability to deposit DNA strands into the nucleus in response to an external stimulus.

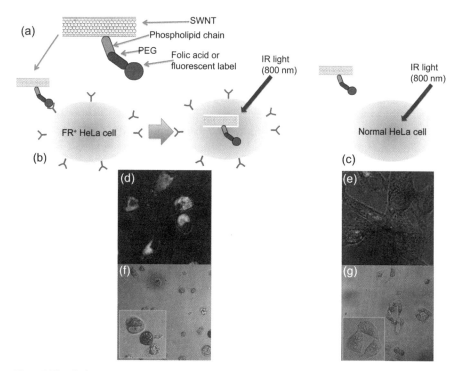

Fig. 6.23 Infrared hyperthermia with carbon nanotubes. (a) Schematic of the nanovector constructed on a single-walled carbon nanotube (SWNT). Short sections (150 nm) of SWNT were rendered soluble and targeted by conjugating with a phospholipid chain attached to a PEG molecule (Section 6.1.4) and terminated by either folic acid or a fluorescent marker. (b) HeLa cells grown in a culture and starved of folic acid, which stimulates them to over-express folic acid receptors on the surface (FR$^+$ cells). The targeted nanotubes bind to the receptors and are then internalized by endocytosis where they absorb infrared light and heat the cell (NB the SWNT nanovector and cell are not to scale. On the scale in the figure the nanovector would be about the size of a full stop and large numbers of them are internalized). (c) The nanovectors are not internalized into normal HeLa cells. (d) confocal microscope image of Fr$^+$ cells incubated with the nanovectors containing both the folic acid and the fluorescent marker cargoes. It is evident that they are inside the cells and clustered around the cell nucleus. (e) Normal cells do not take up significant numbers of the nanovectors. (f) Image of FR$^+$ cells after incubating with the folic acid targeted nanovectors and exposure to 808 nm laser light at 1.4W/cm^2 for 2 minutes. The death of the cells is indicated by their rounded morphology, which is revealed more clearly in the inset. The cells are clearly unable to maintain their normal shape. (g) Normal HeLa cells incubated with the same nanovectors and exposed to infrared light with the same conditions as in (f) remain healthy. Reproduced with the permission of the National Academy of Sciences, U.S.A. from N. Wong et al. [30].

6.3 MEDICAL DIAGNOSIS AND "THERANOSTICS" USING NANOMATERIALS

Increasingly, nanomaterials are being used as probes to report on the position of specific targets within living systems. Using the targeting technology described in Section 6.1.5, nanoparticles or nanotubes can be introduced from outside the body to find and attach to specific types of cell. If their position can be detected from outside the body, they can provide a map of the targeted cells. In addition, some of the nanomaterials that are used to image can also treat a diseased region by either (a) optical or magnetic hyperthermia (Section 6.2) or (b) dumping drugs into the region or directly into the cells. Thus the same agents can be used to detect a diseased region and then provide the treatment. A successful treatment by the agents at the microscopic scale will kill the local diseased cells, which will disperse their contents, including the attached nano-materials, away from the region and eventually out of the body so that they will no longer be detected. Thus suitable nanomaterials could be used to provide an image of, for example, a tumor and when they are in place activated by an external influence (infrared light, oscillating magnetic field, etc.) to provide therapy. On re-imaging the region, any remaining nanomaterials will then report on how effective the treatment has been. This combination of diagnosis and therapy has been labeled "theranostics." Below is a description of how internalized nanomaterials can be detected with position sensitivity from outside the body.

6.3.1 Magnetic Resonance Imaging (MRI) and Contrast Enhancement Using Magnetic Nanoparticles

Magnetic resonance imaging (MRI) has become an important diagnostic tool in medicine due to its ability to obtain contrast between different soft tissues and its good spatial resolution. It is based on the fact that certain atomic nuclei (hydrogen is the relevant element for MRI) possess a permanent magnetic moment and can be considered to be tiny bar magnets. A nucleus, however, is a quantum particle, so in the case of a hydrogen nucleus (a proton) the direction of magnetization is quantized to have just two directions—that is, parallel or antiparallel to an applied field. So if a magnetic field is applied to a large assembly of protons—for example, water molecules within soft tissue in the body—they separate into two populations that are aligned along or against the applied field. These populations have slightly different energies, with the energy separation depending on the strength of the applied field. The energy splitting is very small ($\sim 10^{-7}$ eV), corresponding to the energy of a single photon at radio frequencies, and is much smaller than the thermal energy per particle. For this reason the entire assembly of protons does not fall into the lower energy state, aligned with the field. In fact, so many are thermally excited into the higher energy state that the two populations are almost equal with just a tiny imbalance in the lower energy state. This imbalance produces the so-called longitudinal magnetization. One can modify the state of the two populations by applying a magnetic field pulse that oscillates at the resonant frequency. This can have two effects; that is, it can equalize or even reverse the two populations, thus removing or reversing the longitudinal magnetization, and it can introduce a transverse component to the magnetization (transverse magnetization). The feature that gives MRI its tissue contrast is that the rate at which the longitudinal or transverse magnetization returns to its initial values after applying the pulse is highly sensitive to the precise chemical environment in which the hydrogen nuclei reside. The characteristic time associated with longitudinal relaxation is called T1 and is typically about 1 sec, while for transverse relaxation it is known as T2 and is typically a few tens of milliseconds.

Figure 6.24 illustrates the basic processes in MRI. If one samples the longitudinal or transverse nuclear magnetization a fixed time after each pulse, the different T1 and T2 values for the different types of tissue will give a different signal. The spatial resolution is obtained by introducing a magnetic field gradient through the body so that only one position is in resonance with the oscillating field pulse. Thus as the gradient field is scanned it is possible to assign a T1 or T2 signal with each point in the body and this is the image that is plotted in MRI scans such as that shown in Fig. 6.24f. Whether it is the T1 or T2 signal that is imaged is determined by the timing between the excitation pulse and detection; that is, a long gap ($\sim$1 sec) images using T1 whereas a short gap ($\sim$0.01 sec) images using T2. Just after the excitation pulse the longitudinal magnetization is zero, returning to its pre-pulse value in a characteristic time T1; and the transverse magnetization is maximum, returning to zero after a characteristic time T2. Thus a T1-weighted image records a positive signal relative to background and appears white, while a T2-weighted image records a reduction in signal relative to background and appears dark.

Magnetic compounds introduced by injection have been used for some time to enhance the contrast in MRI images, but recently superparamagnetic iron oxide (SPIO) nanoparticles have become available commercially—for example, Resovist® manufactured by Schering AG—as MRI contrast agents. The nanoparticles cause a significant reduction in T1 and T2 relaxation times in their immediate vicinity, so they show up as white or black on T1-weighted or T2-weighted images, respectively. The available SPIOs are untargeted, so contrast between tumors and healthy tissue is obtained by natural responses of the body to nanomaterials (Section 6.1.4). For example, the liver is normally rich in macrophages known as Kupffer cells, which quickly (5–10 mins) engulf SPIOs. Tumors are devoid of Kupffer cells, so nanoparticles injected into the liver tend to remain in the healthy tissue and a T2-weighted image will show the healthy tissue darkened and any tumors will be light in contrast. An example of contrast enhancement by Resovist® particles is shown in Fig. 6.25 displaying T2-weighted MRI images of a tumor-bearing liver in a patient. The nanoparticle contrast agent darkens the healthy tissue, thereby highlighting the light-colored tumor and also revealing a satellite tumor that is easily overlooked in the image taken without the nanoparticles.

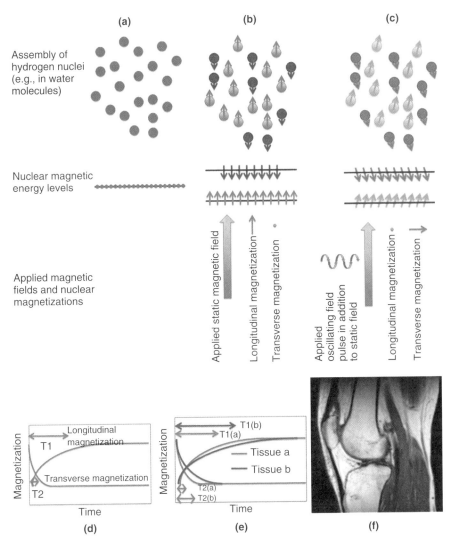

Fig. 6.24 Magnetic Resonance Imaging (MRI). (a) MRI images using hydrogen nuclei (protons) in, for example, proteins or water. The hydrogen nuclei have a permanent magnetic moment that is not apparent till a magnetic field is applied. (b) On application of a magnetic field the protons separate into two populations with their magnetization parallel or antiparallel to the field. The two states correspond to different energies and a slight imbalance between the populations in favor of the parallel alignment produces a small 'longitudinal magnetization' along the field. (c) an additional oscillating field pulse whose frequency is resonant with the energy difference between the populations can equalize or even invert the populations (thus zeroing or inverting the longitudinal magnetization) and also generates a transverse component to the magnetization. (d) After the pulse the longitudinal magnetization relaxes back to its initial value in a characteristic time T1 (typically ~1 s) and the transverse magnetization relaxes back to zero in a time T2 (typically ~0.01 s). (e) The values of T1 and T2 depend sensitively on the chemical environment around the protons and this is used to generate contrast between different types of tissue. (f) MRI image of the knee. Part (f) reproduced from Wikipedia web page: http://en.wikipedia.org/wiki/Mri.

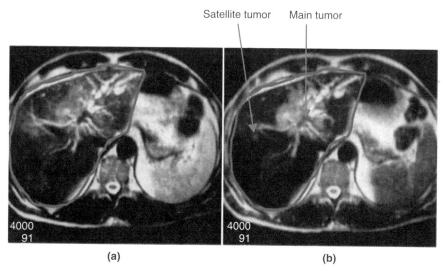

Fig. 6.25 MRI contrast enhancement by Superparamagnetic Iron Oxide (SPIO) nanoparticles. T2-weighted MRI images of a tumor-bearing liver (outlined in red). (a) is taken prior to administering SPIO nanoparticles (Resovist®) and (b) is taken after. The nanoparticles are quickly taken up by macrophages (Kupffer cells), which mostly reside in the healthy tissue. After introducing the contrast agent the healthy tissue is darkened relative to the tumor and the main tumor is much better defined. In addition a small satellite tumor, which could have been missed in (a) becomes clearly visible. Reprinted with permission from the Radiological Society of North America from A. Blakeborough et al. [32].

There are significant improvements in MRI contrast agents expected to emerge in both their selectivity and contrast performance. If active biological targeting (Section 6.1.5) is used, as opposed to relying on the natural response of the body to concentrate the contrast agents in the required area, specific types of cell can be highlighted. In addition, there is some way to go in improving the magnitude of contrast enhancement produced. Contrast is enhanced by the magnetic nanoparticles locally reducing the relaxation time T1 or T2, and it is easy to predict that the higher the magnetization per atom within the particle, the larger the contrast. This has been borne out by recent studies testing the MRI contrast provided by different types of magnetic nanoparticle [33]. A new range of magnetic nanoparticles consisting of ferrites (Fe_2O_4) doped with atoms of Mn, Fe, Co, and Ni were produced chemically (Fig. 6.26a), each compound giving a different average magnetic moment per atom but all higher than the available Fe_2O_3 oxide nanoparticles. Thus it was possible to determine directly the effect of the magnetic moment on MRI contrast. Figure 6.26b shows the relaxivity, R2 (= 1/T2), which is a direct measure of the contrast enhancement, for the various nanoparticles as a function of their magnetic moment/atom. It is evident

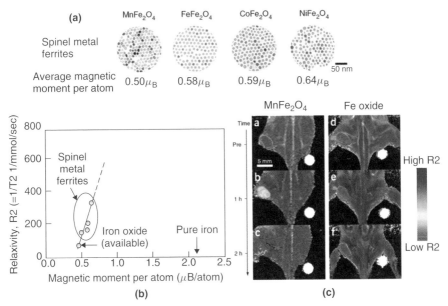

Fig. 6.26 MRI contrast enhancement by spinel ferrite nanoparticles. (a) TEM images of the different types of ferrite nanoparticles used in the study in [33] along with their average magnetic moment per atom. (b) The relaxivity, R2 (=1/T2), which is a direct measure of the strength of contrast enhancement, of the various nanoparticles as a function of their average magnetic moment per atom. Also shown are currently available iron oxide nanoparticles. The magnetic moment of pure Fe nanoparticles (unavailable at present) is indicated showing there are still large gains in R2 achievable. (c) Demonstration of the targeting of $MnFe_2O_4$ nanoparticles conjugated with Herceptin. The images are MRI scans of a mouse bearing a NIH3T6.7 tumor, which is targeted by Herceptin, in its flank with the color map indicating the R2 level within the tumor. The $MnFe_2O_4$ (ferrite) particles injected into the mouse find the tumor and produce a strong contrast enhancement as indicated by the change in color. In contrast iron oxide nanoparticles, conjugated with Herceptin produce very little contrast change. Reproduced with the permission of the Nature Publishing Group from J.-H. Lee et al. [33].

that there is a sharp rise in R2 as the moment increases, but it is also clear from the indicated moment of pure Fe nanoparticles that there are yet larger gains to be found. Pure Fe nanoparticles are not yet available, and for biological use they would need to be coated with a biocompatible shell as discussed in Section 6.2.2.

The nanoparticles used in this study were attached to herceptin, for which a receptor is overexpressed on the cell surfaces of the so-called NIH3T6.7 cancer, which is commonly used in research. Figure 6.26c shows MRI scans at different times of a mouse bearing an NIH3T6.7 tumor after injection of $MnFe_2O_4$–herceptin (left side) and Fe_2O_3–herceptin (right side) nanoparticles. The volume of the tumor is color-coded according to the MRI contrast (R2), and it is clear that the ferrite nanoparticles find the tumor and produce a marked increase in contrast. By comparison the Fe_2O_3–herceptin particles produce very small changes in contrast.

It is a convenient coincidence that the property that gives magnetic nanoparticles a high MRI contrast—that is, an enhanced magnetic moment—is the same one that gives them a high performance in hyperthermia (Section 6.2.2). Thus magnetic nanoparticles could be the ideal theranostic agents discussed at the start of Section 6.3 in that they could target a tumor and reveal it in an MRI scan with high sensitivity, and then a high-frequency magnetic field could be used to generate sufficient localized heat within the nanoparticles to kill the tumor. A following MRI scan would report on how effective the treatment had been, since the only nanoparticles that would be left in place would be those that were still attached to living cancer cells.

6.3.2 Imaging Using Gold Nanoparticles

Gold nanoparticles were described in Section 6.2.3 in the context of optical hyperthermia using the surface plasmon resonance (SPR; see Fig. 6.17), which produces a very strong extinction by the particles for light of the same frequency. The same phenomenon can lead to a very strong scattering of light by particles of a suitable size. For example, Fig. 6.20 shows that for gold particles with a diameter of 80 nm, about half of the total extinction is due to scattering. A gold nanoparticle illuminated with light at the frequency of its SPR will emit up to 100,000 times more photons than, for example, fluorescein, a fluorescent molecule commonly used in imaging, excited by the same light intensity. Although they cannot be resolved in an optical microscope, individual gold nanoparticles can scatter enough light to be visible as a point of light in a microscope image.

Because of the high absorption of visible light by tissue, the use of optical markers is limited to near-surface *in vivo* applications or to *in vitro* research or screening. They are highly convenient probes, however, since immediate visual images of their locations can be obtained in a microscope whereas, for example, imaging the location of magnetic nanoparticles requires complex equipment such as an MRI scanner. As with all nanoparticles, their full potential is realized when they are combined with biological targeting (Section 6.1.5) that locates specific types of cell. Figure 6.27 shows confocal microscope images of Au nanoparticles that have been targeted onto epithelial cervical cancer cells. The types of cell (called SiHa) used in this study [34] overexpress a protein known as epithelial growth factor receptor (EGFR) on their surface membrane, and the gold nanoparticles were attached to anti-EGFR antibodies that bind to this receptor. The image shown in Fig. 6.27a is taken with the microscope focus set to the plane intercepting the top surfaces of the cells, and 647-nm wavelength laser light scattered by the particles is false-colored in red. The scattered light in Fig. 6.27a is overlaid with transmittance images showing the positions

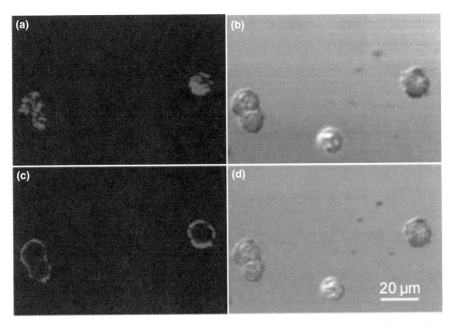

Fig. 6.27   Identifying cervical epithelial cancer cells using Au nanoparticles. (a) Confocal microscope image of cervical epithelial cancer cells labelled with Au nanoparticles conjugated with anti-EFGR antibodies, which target EGFR receptors over expressed on the cancer cell membranes. The microscope focal plane is at the top surface of the cells. (b) the image in (a) overlaid with a transmittance image showing the positions of the cells. Note the unlabelled healthy cell at the bottom center of the image. (c) and (d) as (a) and (b) but with the microscope focus in (c) set to the midplane of the cells. It is clear that the Au nanoparticles are covering the cell membrane of the cancer cells. Reprinted with permission from [34] Fig. 2. © 2003 American Association for Cancer Research.

of the cells in Fig. 6.27b. The two images in Figs. 6.27c and 6.27d are similar but with the microscope focal plane running through the middle of the cells. It is clear that the gold nanoparticles are encasing the cancer cell membranes.

6.3.3   Imaging Using Quantum Dots

Quantum dots, described in detail in Chapter 5, Section 5.2, are usually nanoparticles of semiconducting materials with sizes around 5 nm that can be considered to be "artificial atoms" in which the electronic energy levels are spaced appropriately to emit light. An energy level diagram is shown schematically in Fig. 6.28a, which also illustrates how the quantum dot fluoresces at a fixed wavelength. Short-wavelength light can excite electrons from the occupied to the unoccupied states, and any photon energy greater than the bandgap energy can be used.

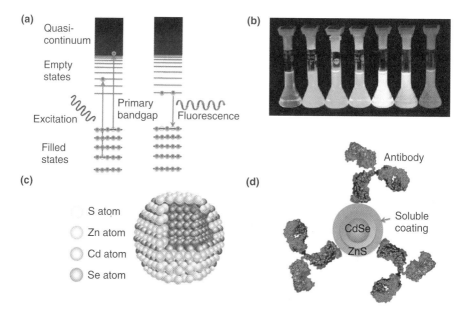

Fig. 6.28 Quantum dots for life science applications. (a) Schematic energy level diagram of quantum dot showing single wavelength fluorescence for a broad range of excitation wavelengths. (b) The primary bandgap can be tuned by changing the size of the quantum dots as shown for a range of CdS/ZnS quantum dot suspensions manufactured by Evident Technologies Inc. (image reproduced with permission of Evident Technologies Inc.). (c) Core-shell structure of an efficient quantum dot with a CdSe core and a ZnS shell. This structure significantly improves the brightness and stability of the dot (see text). (Image reproduced with permission of Evident Technologies Inc.). (d) A fully functionalized quantum dot nanovector complete with a shell that makes the dot soluble and targeting antibodies.

The excited electrons and unoccupied holes quickly relax via nonradiative processes to states on opposite sides of the primary bandgap, in which the lifetime of the excited electrons is relatively long. Thus when recombination occurs, the energy given to the fluorescent photon is fixed. The size of the primary gap and thus the color of the fluorescent photons are easily controlled by changing the size of the quantum dot (Fig. 6.28b). Although fluorescent molecular markers have been used in biological and medical research for a long time, quantum dots have several advantages. They have a much higher quantum yield (number of fluorescent photons out/number of excitation photons in) and they do not suffer from "bleaching," a process where, after emitting a number of photons, a fluorophore permanently loses the ability to fluoresce due to induced chemical changes. In addition, a broad range of excitation photon energies can be used so that in a sample containing quantum dots of different sizes, they can all be excited by the same source and then each emit at their own wavelength. Thus, as shown below, one can simultaneously highlight different parts of a cell by targeting quantum dots of different colors to specific cell structures.

For medical applications a common semiconductor material used is the compound CdSe, illustrated in Fig. 6.28c. The naked CdSe nanoparticle, however, produces relatively weak and unstable emission. This is a common feature of all quantum dots and is due to the fact that the terminated crystal structure at the surface of the dot is unstable against a restructuring of the surface atoms into a new atomic arrangement as described in Chapter 4, Section 4.4.1. The new structure introduces new electronic states (known as traps) that promote the relaxation of the excited state via nonradiative transitions. A solution is to coat the CdSe with a shell of ZnS (Fig. 6.28c), which has a similar crystal structure but a wider bandgap. The shell is not fluorescent; but it enables the CdSe core to maintain its crystal structure across the interface without restructuring, so the good quantum dot behavior of the CdSe dot is restored.

For life science applications, two more shells are required. One is to produce water-soluble suspensions of the quantum dots, and this can be a hydrophobic/hydrophilic layer such as a phospholipid membrane (Section 6.1.2). The coating further passivates the quantum dot and allows bonding to the final layer—that is, targeting molecules that will find specific sites within cells. Figure 6.28d illustrates a fully functionalized quantum dot nanovector fitted with an antibody for targeting.

The power of quantum dots in labeling multiple cell structures is nicely demonstrated in Fig. 6.29, which shows quantum dots fluorescing at 630 nm (red) targeted to the nucleus and at 535 nm (green) targeted to microtubules in a single 3T3 cell, a line originally cultured from mouse embryos [35].

As discussed above, a problem with the external imaging of light emitting nanoparticles *in vivo* is the strong absorption of visible light by tissue, making it difficult to detect them below the skin. Despite this, some work on near-surface *in vivo* imaging has been carried out—for example, the image obtained from the vasculature around mouse ovaries shown in Fig. 6.30 obtained after a tail vein injection of untargeted quantum dots [36]. The fluorescence in this image is being detected through 250 μm of adipose tissue.

Since quantum dots can be made to emit in the red/near-infrared window, where tissue absorption is reduced, their emission can in principle be picked up from deeper layers. A more fundamental problem, however, is that in order for them to fluoresce, they first have to be excited by shorter-wavelength photons that are strongly absorbed. If applied from outside the body, the intensity required to excite deeper-lying quantum dots produces a large background signal and also fluorescence by naturally occurring chromophores in the body swamping the emission by the quantum dots.

A recent innovative approach to get around the problem of *in vivo* imaging using quantum dots uses bioluminescence within the body to excite the

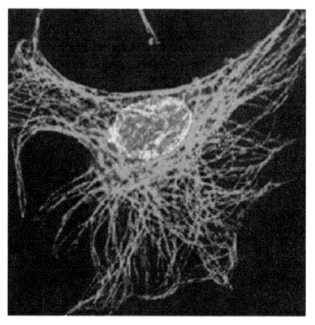

Fig. 6.29 Multiple color labeling by quantum dots. A single 3T3 cell labeled by quantum dots fluorescing at two wavelengths, that is, 630 nm (red) and 535 nm (green). The red dots have been targeted by anti-nuclear antigen to the nucleus and the green dots by mouse anti-a-tubulin antibody to the microtubules. Reproduced with the permission of the Nature Publishing Group from X. Wu et al. [35].

quantum dots internally without the need for an external excitation source [37]. Bioluminescence, whereby the energy of a chemical reaction is used to produce light, has evolved independently a number of times in animals, plants, insects, and bacteria and is particularly prevalent in marine life. For example, the sea pansy (*Renilla reniformis*) uses a pigment-containing the protein, *R. reniformis* luciferase, to catalyze a reaction between coelenterazine and oxygen to produce light and CO_2 as by-products. This type of reaction can be extremely efficient, converting more than 90% of chemical energy to light.

The effectiveness of bioluminescence to excite quantum dots *in vivo* was tested by conjugating polymer-coated CdSe/ZnS quantum dots sized to emit at 655 nm (red) to a variant of R. *reniformis* luciferase known as Luc8 (Fig. 6.31a). On exposing the quantum dot conjugate to coelenterazine, blue light with a wavelength of 480 nm emitted in the reaction excites the quantum dot and stimulated its emission at 655 nm (Fig.6.31b) as shown in the emission spectrum in Fig. 6.31c. The conjugates were tested in mice by injecting the QD655–Luc8 conjugates and just Luc8 for comparison at different depths and measuring the

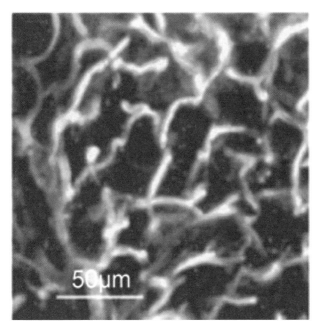

Fig. 6.30 In vivo imaging by quantum dots. The vasculature around mouse ovaries obtained using untargeted quantum dots introduce by tail vein injection. The image has been obtained through 250 μm of adipose tissue. Reproduced with the permission of the American Association for the Advancement of Science (AAAS) from D. R. Larson et al. [36].

light emission from the injection sites after the introduction of coelenterazine by a tail vein injection. The results are shown in Fig. 6.31d, indicating the four injection sites. The QD655–Luc8 conjugate was injected at sites I and III on the left side of the mouse, with site I being just under the skin (subcutaneous) while site III was intramuscular at 3 mm below the surface. For comparison, sites II and IV were at a depth similar to that of sites I and III, respectively, but the injection was Luc8 only. The photon count is color-coded according to the scale in the image, and it is observed that while the subcutaneous sites both show strong emission after the addition of the Coelenterazine, the intramuscular emission is much stronger from site III containing the QD655–Luc8 conjugate. This demonstrates that the blue emission from Luc8 is much more strongly absorbed by the intervening tissue than is the red light from the quantum dots. It is also clear from the study that utilizing bioluminescence to internally stimulate quantum dots is a viable method and greatly enhances the prospects for *in vivo* imaging using quantum dots. The quantum dots used can be made to emit in the infrared region, further increasing the viable depth for imaging.

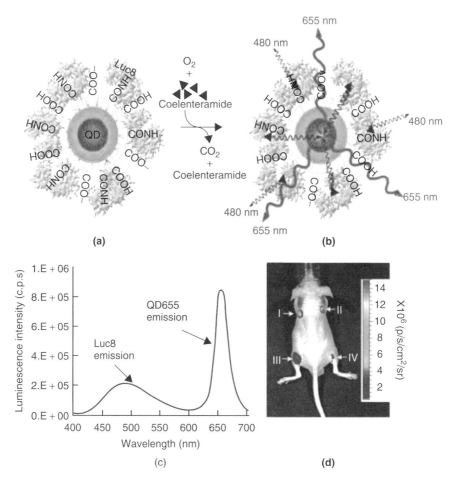

Fig. 6.31 Using bioluminescence to excite quantum dots in vivo. (a) Polymer-coated CdSe/ZnS quantum dots conjugated with Luc8, a variant of the protein R. reniformis luciferase used by the sea pansy (renilla reniformis) for bioluminescence. (b) On mixing with Coelenterazine Lu8 catalyses a reaction with oxygen to produce light with a wavelength of 480 nm (blue). This emission excites the quantum dot to emit at 655 nm (red). (c) The emission spectrum from QD655-Luc8 conjugates after adding Coelenterazine showing emission at both frequencies. (d) Injection sites in a mouse for QD655-Luc8 conjugates and Luc8 alone for comparison. Sites I and II were injections just under the skin (subcutaneous) while sites III and IV were intramuscular 3 mm below the surface. The injections on the left side of the mouse (sites I and III) were of QD655-Luc8 conjugates while those on the right side (sites II and IV) were Luc8 alone. The photon emission intensity from the four sites after adding Coelenterazine by a tail vein injection is shown by the color scale. While strong emission occurs from both subcutaneous sites the emission from the intramuscular sites is much stronger from site III containing the QD655-Luc8 conjugate than site IV containing Luc8 alone. This is due to the stronger absorption of blue light by the intervening tissue. Reproduced with the permission of the Nature Publishing Group from M-K. So et al. [37].

PROBLEMS

1. A nanovector consists of a nanoparticle with a pure Fe core of diameter 50 nm coated with a biocompatible shell of thickness 10 nm. Calculate the steady velocity of the particle in still blood in an applied magnetic field gradient of 50 T/m. The density and atomic mass of Fe are 7860 kg/m^3 and 55.85 daltons. All other parameters are given in Advanced Reading Box 6.1.

2. During magnetic nanoparticle hyperthermia, in steady-state conditions, the temperature in the heated region containing the nanoparticles stabilizes at a constant temperature typically 10°C above the core body temperature. Moving away from the nanoparticle filled region, the temperature decays back to the core body temperature exponentially with distance as shown in Advanced Reading Box 6.2. Compare the characteristic decay distances in muscle and in fat using the parameters in Table 6.1 and in Advanced Reading Box 6.2, assuming the blood perfusion rates are the same.

3. The figure below shows a measurement of the imaginary part of the magnetic susceptibility of a suspension of superparamagnetic magnetic nanoparticles used for hyperthermia. The total mass of magnetic material in the suspension is 10 mg. Calculate the specific absorption rate (SAR) of the nanoparticles for an applied alternating magnetic field with a frequency of 200 kHz and an amplitude of 2500 A/m. It appears that the particles may perform better at a frequency of about 1 MHz, Why would it be disadvantageous to use the higher frequency in practice?

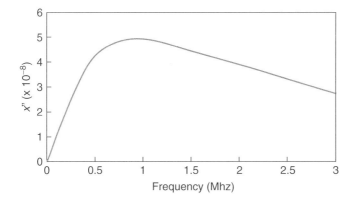

4. The figure below shows the real part of the dielectric constant of gold as a function of wavelength in the visible region of the spectrum. Estimate the wavelength of the surface plasmon resonance (SPR) in Au nanoparticles whose diameter is very much smaller than the wavelength of visible light immersed in water and immersed in blood using the parameters given in Advanced

Reading Box 6.5. In principle the SPR can absorb sufficient power to be used for hyperthermia, but tissue is highly absorbing at visible light wavelengths so it would not work at depths greater than a few millimeters below the skin. How could the wavelength of the SPR be shifted into the near-infrared region of the spectrum at which tissue is relatively transparent?

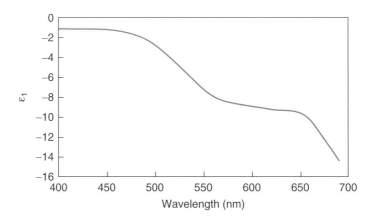

5. Describe how magnetic nanoparticles could be used at the core of "theranostic" nanovectors that could be used for both diagnosis and treatment of cancer.

REFERENCES

1. *Cancer Facts and Figures* 2007, American Cancer Society, Atlanta, 2007.

2. Q. Zeng, I. Baker, J. A. Loudis, Y. Liao, P. J. Hoopes, and J. B. Weaver, Fe/Fe oxide nanocomposite particles with large specific absorption rate for hyperthermia, *Appl. Phys. Lett.* **90** (2007), 233112.

3. A. S. Zahr, C. A. Davis, and M. V. Pishko, Macrophage uptake of core-shell nanoparticles surface modified with poly(ethylene glycol), *Langmuir* **22** (2006), 8178–8185.

4. O. C. Farokhzad, J. Cheng, B. A. Teply, I. Sherifi, S. Jon, P. W. Kantoff, J. P. Richie, and R. Langer, *Proc. Natl. Acad. Sci.* **103** (2006), 6315–6320.

5. J. F. Kukowska-Latallo, K. A. Candido, Z. Cao, S. S. Nigavekar, I. J. Majoros, T. P. Thomas, L. P. Balogh, M. K. Khan, and J. R. Baker, Nanoparticle targeting of anticancer drug improves therapeutic response in animal model of human epithelial cancer, *Cancer Res.* **65** (2005), 5317.

6. K. Halbach, Physical and optical properties of rare earth cobalt magnets," *Nuclear Instrum. Methods* **187** (1981), 109–117.

7. S.–I. Takeda, F. Mishima, S. Fujimoto, Y. Izumi, and S. Nishijima, *J. Magn. Magn. Mater.* **311** (2007), 367–371.

8. P. Dames, B. Gleich, A. Flemmer, K. Hajek, N. Seidl, F. Wiekhorst, D. Eberbeck, I. Bittmann, C. Bergemann, T. Weyh, L. Trahms, J. Rosenecker, and C. Rudolph,

Targeted delivery of magnetic aerosol droplets to the lung, *Nature Nanotechnol*. **2** (2007), 495–499.

9. W. B. Coley, A preliminary note on the treatment of inoperable sarcoma by the toxic product of erysipelas, *Post-graduate* **8** (1893), 278–286; W. Coley, The treatment of malignant tumours by repeated innoculations of erysipelas: With a report of ten original cases, *Am. J. Med. Sci*. **105** (1893), 487–510.

10. J. Overgaard, D. Gonzalez Gonzalez, M. C. C. H. Hulshof, G. Arcangeli, O. Dahl, O. Mella, and S. M. Bentzen, Hyperthermia as an adjuvant to radiation therapy of recurrent or metastatic malignant melanoma: A multicentre randomized trial by the European Society for Hyperthermic Oncology, *Int. J. Hyperthermia* **12** (1996), 3–20.

11. H. Pennes, Analysis of tissue and arterial blood temperatures in the resting human forearm, *J. Appl. Physiol*. **1** (1948), 93–122.

12. R. K. Gilchrist, R. Medal, W. D. Shorey, R. C. Hanselman, J. C. Parrott and C. B. Taylor, Selective inductive heating of lymph nodes, *Ann. Surgery* **146** (1957), 596–606.

13. R. E. Rosenweig, Heating magnetic fluid with alternating magnetic field, *J. Magn. and Magn. Mater*. **252** (2002), 370–374.

14. S. Mornet, S. Vasseur, F. Grasset, and E. Duguet, Magnetic nanoparticle design for medical diagnosis and therapy, *J. Mater. Chem*. **14** (2004), 2161–2175.

15. R. Hergt, W. Andrä, C. G. d'Ambly, I. Hilger, W. A. Kaiser, U. Richter, and H. G. Schmidt, *IEEE Trans. Magn*. **34** (1998), 3745–3754.

16. A. Jordan, T. Rheinländer, N. Waldöfner, and R. Scholz, *J. Nanoparticle Res*. **5** (2003), 597–600.

17. N. A. Brusentsov, V. V. Gogosov, T. N. Brusentsova, A. V. Sergeev, N. Y. Jurchenko, A. A. Kuznetsov, and L. I. Shumakov, *J. Magn. Magn. Mater*. **225** (2001), 113–117.

18. N. A. Brusentsov, L. V. Nikitin, T. N. Brusentsova, A. A. Kuznetsov, F. S. Bayburtskiy, L. I. Shumakov, and N. Y. Jurchenko, *J. Magn. Magn. Mater*. **252** (2002), 378–380.

19. M. Johanssen, U. Gneveckow, K. Taymoorian, B. Thiesen, N. Waldöfner, R. Scholz, K. Jung, A. Jordan, P. Wust, and S. A. Loening, Morbidity and quality of life during thermotherapy using magnetic nanoparticles in locally recurrent prostate cancer: Results of a prospective phase I trial, *Int. J. Hyperthermia* **23** (2007), 315–323.

20. K. Maier-Hauff, R. Rothe, R. Scholz, U. Gneveckow, P. Wust, B. Thiesen, A. Feussner, A. V. Deimling, N. Waldöfner, R. Felix, and A. Jordan, Intercranial thermotherapy using magnetic nanoparticles combined with external beam radiotherapy: Results of a feasibility study on patients with *Glioblastoma multiforme*, *J. Neurooncol*. **81** (2007), 53–60.

21. C. F. Bohren and D. R. Huffman, *Absorption and Scattering of Light by Small Particles*, John Wiley & Sons, New York, 1983.

22. E. D. Palik, *Handbook of Optical Constants*, Academic Press, Orlando, FL, 1985.

23. P. K. Jain, K. S. Lee, I. H. El-Sayed, and M. A. El-Sayed, Calculated absorption and scattering properties of gold nanoparticles of different size, shape, and composition: Applications in biological imaging and biomedicine, *J. Phys. Chem. B* **110** (2006), 7238–7248.

24. S. Link and M. A. El-Sayed, Size and temperature dependence of the plasmon absorption of colloidal gold nanoparticles, *J. Phys. Chem. B* **103** (1999), 4212–4217.

25. J. Eichler, J. Knof and H. Lenz, Measurements on the depth of penetration of light (0.35–1.0 μm) in tissue, *Radiat. Environ. Biophys*. **14** (1977), 239–242.

26. C. Loo, A. Lin, L. Hirsch, M.-H. Lee, J. Barton, N. Halas, J. West, and R. Drezek, Nanoshell-enabled photonics-based imaging and therapy of cancer, *Technol. Cancer Res. Treatment* **3** (2004), 33–40.

27. X. Huang, I. H. El-Sayed, W. Qian, and M. A. El-Sayed, Cancer cell imaging and photothermal therapy in the nearinfrared region by using gold nanorods, *J. Am. Chem. Soc*. **128** (2006), 2115–2120.

28. L. M. Liz-Marzan, *Mater. Today* **7** (2004), 26.

29. P. K. Jain, I. H. El-Sayed, and M. A. El-Sayed, Au nanoparticles target cancer, *Nanotoday* **2** (2007), 16–29.

30. N. Wong Shi Kam, M. O'Connell, J. A. Wisdom, and H. Dai, Carbon nanotubes as multifunctional biological transporters and near-infrared agents for selective cancer cell destruction, *Proc. Natl. Acad. Sci*. **102** (2005), 11601.

31. C. J. Gannon, P. Cherukuri, B. I. Yakobson L. Cognet J. S. Kanzius, C. Kittrell, R. B. Weisman, M. Pasquali, H. K. Schmidt, R. E. Smalley, and S. A. Curley, Carbon nanotube-enhanced thermal destruction of cancer cells in a noninvasive radiofrequency field, *Cancer* **110** (2007), 2654–2665.

32. A. Blakeborough, J. Ward, D. Wilson, M. Griffiths, Y. Kajiya, J. A. Guthrie, and P. J. A. Robinson, Hepatic lesion detection at MR imaging: A comparative study with four sequences, *Radiology* **203** (1997), 759–765.

33. J.-H. Lee, Y.-M. Huh, Y.-W. Jun, J.-W. Seo, J.-T. Jang, H.-T. Song, S. Kim, E.-J. Cho, H.-G. Yoon, J.-S. Suh, and J. Cheon, Artificially engineered magnetic nanoparticles for ultra-sensitive molecular imaging, *Nature Med*. **13** (2007), 95–99.

34. K. Sokolov, M. Follen, J. Aaron, I. Pavlova, A. Malpica, R. Lotan, and R. Richards-Kortum, Real-time vital optical imaging of precancer using anti-epidermal growth factor receptor antibodies conjugated to gold nanoparticles, *Cancer Res*. **63** (2003), 1999–2004.

35. X. Wu, H. Liu, J. Liu, K. N. Haley, J. A. Treadway, J. P. Larson, N. Ge, F. Peale, and M. P. Bruchez, Immunofluorescent labeling of cancer marker Her2 and other cellular targets with semiconductor quantum dots, *Nature Biotechnol*. **21** (2003), 41–46.

36. D. R. Larson, W. R. Zipfel, R. M. Williams, S. W. Clark, M. P. Bruchez, F. W. Wise, and W. W. Webb, Water-soluble quantum dots for multiphoton fluorescence imaging *in vivo*, Science **300** (2003), 1434–1436.

37. M-K. So, C. Xu, A. M. Loening, S. S. Gambhir, and J. Rao, Self-iIlluminating quantum dot conjugates for *in vivo* imaging, *Nature Biotechnol*. **24** (2006), 339–343.

CHAPTER 7

Radical Nanotechnology

The most far-reaching version of nanotechnology, described as *radical nanotechnology*, is the construction of machines whose mechanical components are the size of molecules. Although under a single umbrella, this encompasses technological goals that are often not distinguished but are actually quite different. One aspect of radical nanotechnology is *molecular manufacturing*, which postulates the possibility of mechanically placing atoms at controlled positions to assemble a macroscopic object. There are plenty of examples of the deliberate positioning of atoms and nanoparticles using scanning tunneling microscopes (STMs; see Chapter 4, Section 4.4.2) or atomic force microscopes (AFMs), but these were carried out slowly by a human operator. In addition, one must carefully choose the combination of atoms and surfaces to make this possible. Molecular manufacturing would require a manipulator/assembler, which is itself microscopic, able to recognize the different atoms around it and assemble them. There would also need to be a huge number of copies of the assembler. For example, a mobile phone contains more than 10^{24} atoms; so even if an assembler could position a million atoms per second (which is not unreasonable—things work fast at the nanoscale), it would still take more than 10^{18} seconds or 30 billion years (about the age of the universe) to assemble the phone. Vast numbers of copies are not necessarily a problem, especially if the assembler can build a copy of itself. A microscopic assembler would probably contain less than a billion atoms, so it could build a copy of itself in about 15 minutes. These two could then start replicating, and so on, so the population would double every 15 minutes. After about 5 hours there would be a million and after just 15 hours there would be enough assemblers (10^{18}) to manufacture a phone per second. This is sometimes referred to as exponential molecular manufacturing because of the exponential growth of any self-replicating system.

Clearly, there would have to be a hierarchy of assemblers working at different scales so that the atomic assemblers could pass finished blocks on to block assemblers and after a few steps the assemblers would be micromachines that can be manufactured by top-down methods available today. The entire process chain within a *nanofactory* based on molecular manufacturing is illustrated in a

Introduction to Nanoscience and Nanotechnology, by Chris Binns
Copyright © 2010 John Wiley & Sons, Inc.

video available on YouTube [1] sponsored by Nanorex Inc. Even for the cynical, it is an impressive vision of molecular manufacturing.

The massive unsolved problem with this technology is the very first step—that is, assembling atoms or molecules of the correct type in the right sequence. This has been famously summarized by the late Nobel Laureate Richard Smalley, who called it the "fat fingers" problem and the "sticky fingers" problem [2]. That is, any assembler would have to grip atoms with grippers that were much bigger than the objects they were trying to pick up and would inherently stick to anything they touched. Here there may be some clues from biology because within every living cell there are assemblers (polymerases) that manufacture DNA by assembling nucleic acids in the correct sequence. There are also assemblers (ribosomes) that manufacture proteins by assembling amino acids in the correct sequence. Both of these are able to sample randomly arriving molecules and distinguish with very high fidelity between similar structures. They are even able to correct errors should they occur. These biological assemblers will be discussed in Section 7.4.

The other aspect of radical nanotechnology is the creation of small autonomous robot-like machines (nanobots) that have some on-board processing capability and an energy source so they are able to move, either over a surface, in liquids, or through the air. These could also be equipped with molecular assemblers that could build objects out of atoms in their path including copies of themselves. It was this aspect of the technology that evoked the famous doomsday "gray-goo" scenario originally highlighted by Eric Drexler (see reference 5 in the Introduction) in which the nanobots replicate out of control. It is very easy to continue the exponential calculation above to work out that in only a few days there could be sufficiently vast numbers of nanobots to munch through the Earth and rearrange its atoms into a gray goo.

These days nobody is suggesting building self-replicating nanobots. Radical nanotechnology has condensed into two branches: (a) molecular manufacturing based on the nanofactory (see reference 6 in the Introduction), which does not involve any self-replication, and (b) nonreplicating nanobots that perform a useful task; the earliest envisaged applications are in medicine. The rest of the chapter discusses any enabling technologies for these two branches that move toward, making either of them feasible.

7.1   LOCOMOTION FOR NANOBOTS AND NANOFACTORIES

One thing that we all expect a robot to do is to move, and the nanofactory requires that materials down to the size of molecules are moved from one place to another, so all aspects of radical nanotechnology require movement and methods of transport. We can learn an enormous amount about controlled motion from living organisms that have developed nanoscale walkers, conveyor belts, motors and propellers that can exert remarkable force and function very efficiently. Cells simply rely on random motion for small molecules to travel short distances, but this is far too slow to move larger cargos significant distances across the cell. Thus sophisticated transport mechanisms that move selected cargoes from the nucleus to the outer edge of the cell and in the opposite direction have evolved. One of the best-understood mechanisms involves a protein known as *kinesin* moving along tubular filaments called *microtubules*. These are one of components of the cytoskeleton of a cell (see Chapter 6, Section 6.1.3) and consist of tubes, which are themselves composed of long rods of polymerized proteins known as *protofilaments*. A microtubule has diameter of about 25 nm and a length varying from 200 nm to 25 μm, so in terms of size it is similar to a multi-walled carbon nanotube (MWNT). They perform several tasks including structural support for the cell as part of its cytoskeleton, growing and shrinking to change the cell shape, and correct segregation of chromosomes during cell division.

The microtubule function relevant to this section is that they also act as railway tracks within cells along which the kinesin proteins act as locomotives pulling cargo. There are a number of kinesins, but the generic structure is shown in Fig. 7.1a. It consists of a motor domain, which is the two heads that bind and unbind to the microtubule to produce motion, a long coiled stalk, and a tail region that attaches to the cargo. The mechanism by which kinesin walks along microtubules is a hand-over-hand mechanism [3] in which the back head releases from the microtubule and is thrown forward by the leading head to bind at the next binding site. The process is illustrated in Fig. 7.1b and relies on the two neck regions, which can be attached to their respective head only at the end or can be zippered tightly along the head. The heads attach to one of the two protein subunits (shaded green) along the protofilament of the microtubule. The initial state shown is where the neck on the back head is zippered, and both heads are attached to the microtubule. In the following step the trailing head hydrolyzes adenosine triphosphate (ATP) and releases phosphate that causes it to release from the microtubule and its neck to unzipper. The forward head releases an adenosine diphosphate (ADP) molecule and absorbs an adenosine triphosphate (ATP) molecule, which causes its neck to zipper, throwing the rear head forward. This then binds onto the next protein subunit along the protofilament, and the roles of the two heads reverse. Each cycle moves the kinesin 8 nm along the microtubule and ~100 steps are taken per second moving the cargo at ~0.8 μm sec^{-1}. Thus the microtubule–kinesin system can move cargo from the center to the outside of a cell in a few seconds. An excellent video showing the process in more detail has been created by Ron Milligan at the Scripps Research Institute [4].

Kinesins are members of a general class of motor proteins that convert the chemical energy contained in ATP into the mechanical energy of movement. Another group is dyneins, which also travel along microtubules but in the opposite direction moving cargo from the outer part of the cell to the nucleus.

Since 1985, kinesin has successfully been made to work *in vitro* but with its role inverted in that its tail groups have been immobilized on surfaces and while the heads have been exposed. Microtubules resting on a bed of kinesin heads are moved around rather like crowd surfers. This is illustrated clearly in the images taken by Hess et al. [5] at the University of Washington of microtubules

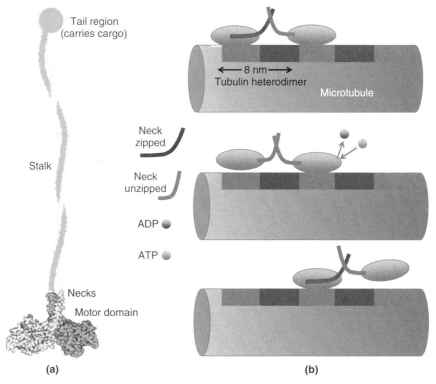

Fig. 7.1 Kinesin transport along microtubule. (a) Generic kinesin structure showing the two heads and necks in the motor domain a long stalk and the tail region onto which the cargo is attached. Reproduced from Wikipedia web page: (http://en.wikipedia.org/wiki/Kinesin) (b) Schematic of translational mechanism. In the top image both heads are attached to protein subunits (green) of the protein heterodimers that compose the protofilaments of the microtubule. The neck of the trailing head is zippered. In the next step (middle image) the trailing head hydrolyzes adenosine triphosphate (ATP) and releases phosphate that causes it to release from the microtubule and its neck to unzipper. The forward head releases an adenosine diphosphate (ADP) molecule and absorbs an adenosine triphosphate (ATP) molecule, which causes its neck to zipper throwing the rear head forward. This then binds onto the next protein subunits along the protofilament and the roles of the two heads reverses.

moving around on a surface coated with kinesin in the presence of obstacles. The substrate was polyurethane, onto which was patterned an array of pillars 10 μm in diameter and 1 μm high. Prior to depositing the kinesin from solution, the surface was coated with casein, a protein that is known to help the kinesin bind to the surface and improves its function. After the kinesin was adsorbed homogeneously over the surface, the microtubules with a diameter of 24 nm and an average length of 1.5 μm were added. They were fluorescently labeled so their position could be tracked with an optical microscope. Without ATP the

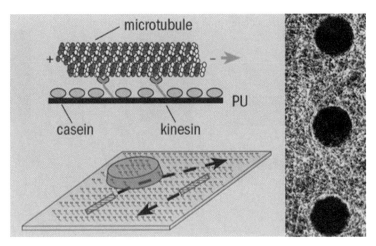

Fig. 7.2   Random motion of crowd-surfing tubules on a kinesin covered surface. Kinesin immobilized by its tail groups on a polyurethane surface pre-coated with casein will move microtubules randomly over the surface. The surface was patterned with pillars 10 μm in diameter and 1 μm high. The right image shows a sum of 500 frames taken every 5 s by an optical microscope of the fluorescently labeled tubules. The entire image is covered by random light spots apart from the pillars, which the microtubules are unable to climb due to their stiffness. Reprinted with permission from [5]. © 2002 by the American Chemical Society.

kinesin is unable to work and the microtubules are stationary, so addition of ATP to the solution "switches on" the motion. Figure 7.2 shows a schematic of the casein/kinesin polyurethane surface and a sum of 500 images, taken every 5 seconds of the microtubule motion after the addition of ATP. The random motion of the tubules has covered the entire frame with light spots, but the pillars remain dark because the microtubules are too stiff to bend upwards sufficiently for the kinesin on the pillar sidewalls to carry them over the pillars.

Clearly, microtubules moving on beds of kinesin would make good conveyor belts in a nanofactory if the random motion could be controlled and more recent work has focused on achieving this. Yuichi Hiratsuka et al. [6] fabricated micro channels for the microtubule conveyors using UV photolithography as shown in Fig. 7.3. This is similar to electron beam lithography (EBL) presented in Chapter 4, Section 4.2.1 except that the resist becomes soluble when exposed to UV light. It is a rapid way of patterning micron-sized features on surfaces. They patterned a series of channels into resist coated on a glass substrate (Figs. 7.3a–7.3c) and then deposited kinesin from solution onto the pattern (Fig. 7.3d). The protein attaches preferentially to glass so that most of the kinesin coats the bottom of the channels (Fig. 7.3e). Finally, the microtubules were added and as in the experiment in Fig. 7.2, they were fluorescently labeled so their positions could be tracked with an optical microscope.

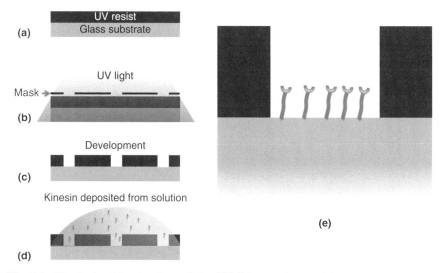

Fig. 7.3 Producing kinesin channels by UV lithography. (a) Glass substrate is coated with resist. (b) Resist is exposed to UV light through a mask containing the required pattern of channels. (c) The resist is developed to etch the pattern of channels down to the glass substrate. (d) the kinesin is deposited from solution onto the patterned surface. (e) Kinesin binds preferentially to the glass substrate coating the bottom of the channels.

As well as channeling the flow, it is important to achieve directional control. There is an inherent directionality in the kinesin/microtubule conveyor because the microtubules are polar and the kinesins generally walk toward a specific end (termed the plus end). In the crowd-surfing experiments, however, the kinesins are randomly oriented as are the microtubules so some external directional control has to be applied. This was achieved in the experiments of Hiratsuka et al. by patterning the arrowhead structure shown in Fig. 7.4a into the channels. A microtubule entering the arrowhead from the wrong direction has a high probability of having its direction reversed and this is confirmed by the series of images Figs. 7.4b–7.4f taken at the times specified in the frames after the ATP was added to "switch on" the system. It is observed that the fluorescent microtubules are restricted to the circular track and selected individuals are artificially colored in the images so that their direction of motion around the track can be observed. Clearly, the arrowheads have imposed a different direction in the two tracks as observed by the direction of evolution of the coloured microtubules.

Further control over the microtubules was demonstrated by van den Heuvel et al. [7], who deposited kinesin in channels whose base was coated with a gold electrode; the lid of the liquid cell also had gold electrodes. By applying an electric field between the electrodes, they found an enhanced tendency of the microtubules in solution to dock with the kinesin heads and start moving. On removing the field, they observed a partial detachment of the microtubules

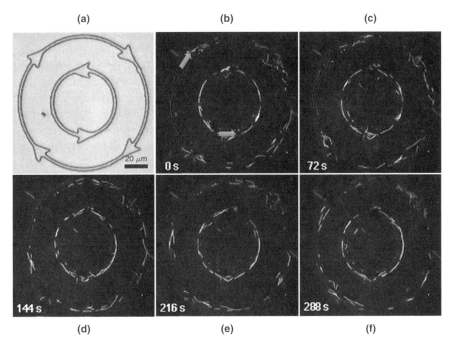

Fig. 7.4   Directional control over motion of microtubules on kinesin. (a) Arrowheads patterned into channels to impose directional flow. A microtubule entering an arrowhead from the wrong direction has a high probability of having its direction reversed. (b)-(f) series of microscope images of fluorescent microtubules taken at the specified time after the addition of ATP. Clearly the microtubules are restricted to the circular channels. Selected individuals have been artificially colored so that their direction of motion can be observed. It is seen that the evolution of the colored tracks is in the direction imposed by the arrows in (a). Reproduced with the permission of Elsevier Science from Y. Hiratsuka et al. [6].

from the kinesin to go back into solution. This affords the ability to create and dismantle the conveyor belt by a simple external control.

The kinesin/microtubule transport system has been emphasized here because it is relatively well understood and techniques for controlling it are quite advanced. Its most foreseeable utilization in radical nanotechnology is to produce conveyor belts for transporting material around the nanofactory. The main remaining problem is to achieve selective attachment of a cargo onto the microtubule carrier and selective removal when required. There are, however, many other types of biological walkers, and recently there has been an increasing focus on those based on DNA [8–11].

DNA walkers are much smaller than kinesin, and at this point it is worth discussing the role of random Brownian motion in nanolocomotive devices. All biological motors work in a liquid environment, which to a nanomachine does not appear to be a continuous fluid but instead a seething mass of water molecules moving randomly at high speed. The devices are sufficiently small that the momentum imparted by water molecule impacts will produce a significant movement so at the nanoscale everything is jiggling about. The random motion was first reported by the botanist Robert Brown in 1827 when he was observing tiny pollen particles in water through a microscope, hence the name. Brownian motion would be a nightmare for an engineer trying to design a high-precision machine, but living systems have evolved in this environment and make good use of the random motion. For large systems like kinesin the random jiggling is quite small, and it is useful for achieving alignment of "lock-and-key"-type fits between different parts of the molecule.

For much smaller walkers like DNA strands, the random motion is quite severe and trying to get a well coordinated deliberate walking action is very difficult. The normal approach is to try and introduce a one-way ratchet effect and let Brownian motion do the rest. Thus the idea is to set up a structure that, when shaken, tends to move in just one direction. It is important to point out that this is not extracting useful work from the random motion of the water molecules, which would contravene the second law of thermodynamics. Molecular ratchet systems are driven by chemical energy, but Brownian motion is used to bring complementary units together. An example of a DNA-based ratchet system devised by Bath et al. at the University of Oxford [12] is shown in Fig. 7.5. The track is a line of short single strands of DNA (see Fig. 7.5), and the cargo (green) is a strand with a complementary set of bases to the track segments. An enzyme is used to cleave the double DNA strand at the point shown so that a short segment at the end of the track strand is lost, leaving a short single strand at the end of the cargo. Brownian motion will enable the exposed end of the cargo strand to find the complementary bases at the end of the next track strand and also for the bottom end of the cargo strand to exchange the current track strand bottom with the neighboring one so that at the end of the cycle the cargo has shifted one track position and there is no mechanism by which it can return. This is an elegant and simple mechanism and has the advantage of a track composed of just one type of DNA strand, but a drawback is that the track is altered by the passage of the cargo. Other more complex DNA walkers have a track composed of a specific sequence of single-stranded DNA segments, and the two legs of the walker are activated by specific "instruction" DNA segments in the solution [13]. These leave the track unscathed, but the track has to be initially installed with a repeating sequence of several different DNA strands.

All the motion devices described so far will execute linear motion along a track immersed in a liquid containing the right chemicals. There is no walker that will travel across a general surface in a controlled way in any direction chosen by some on-board processing capability. There are many types of biological cells that have whip-like tails known as *flagella* (*flagellum* in the singular) that are rotated at high speed to propel the cell through a liquid medium in any direction. Cells fall into two basic classes: (a) *prokaryotes*, which do not have a nuclear membrane (see Chapter 6, Section 6.1.3) and include all bacteria, and (b) *eukaryotes*, which keep their genetic material enclosed within a well-defined nucleus. This latter group

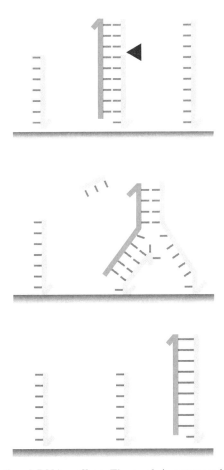

Fig. 7.5   A unidirectional DNA walker. The track is composed of a row of identical short single DNA strands and the cargo (green) is a single DNA strand with a complementary set of bases to the track segments (see Fig. 6.6). An enzyme is used to cleave the double DNA strand at the point shown so that a short segment at the end of the track strand is lost leaving a short single strand at the end of the cargo. Brownian motion will enable the exposed end of the cargo strand to find the complementary bases at the end of the next track strand and also for the bottom end of the cargo strand to exchange the current track strand bottom with the neighboring one so that at the end of the cycle the cargo has shifted one track position and there is no mechanism by which it can return. Reproduced with the permission of the Nature Publishing Group from J. Bath and A. J. Turberfield [9].

includes cells within all plants and animals. The distinction is drawn here because, although both types have developed flagella for propulsion, the driving motors are different in their layout. An additional classification for the bacteria is whether they are *Gram-positive* or *Gram-negative*, which determines the details of their outer membrane structure. This distinction is also significant as the flagellum motor is embedded in the cell membrane, and its configuration is different in the two groups.

Here the focus is on bacterial flagella to highlight the basic operation, and the flagellum and motor of a Gram-positive bacterium are shown schematically in Fig. 7.6. The flagellum itself is composed of a protein called *flagellin*, which naturally forms tubular structures. Its diameter is ~20 nm and lengths can be

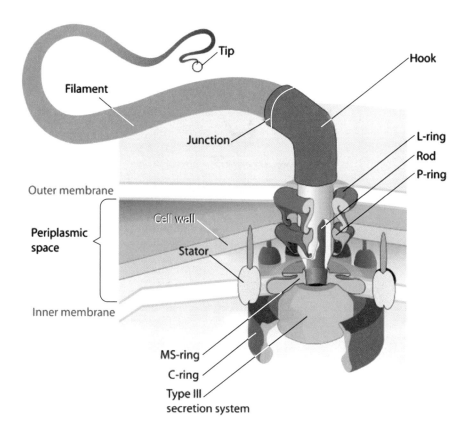

Fig. 7.6 Flagellum and motor of a Gram-negative bacterium. The flagellum is composed of the protein flagellin, which forms tubular structures and has a diameter of ~20 nm and a length of up to 10 μm. The hook near the cell membrane produces a sharp bend in the flagellum so that when it is rotated it moves in a helix to produce thrust. A straight shaft attached to the hook passes through four protein rings (labeled L, P, MS and C) that act as bearings and in addition, the lower two are thought to provide the torque to drive the flagellum [14]. The energy to operate the motor is derived from hydrogen ions (protons) crossing the cell membrane as a result of their electrostatic interaction and concentration gradient. The latter can be controlled by the cell's metabolism to vary the rotational speed of the motor. Reproduced from Wikipedia web page: http://en.wikipedia.org/wiki/Flagellum.

up to 10 μm, which is about the same size as the cell body. At the cell end is a hook that bends it at a sharp angle relative to the cell body, which ensures that it performs a helical motion when it is rotated. The hook is attached to a straight shaft that passes through rings of protein embedded in the cell membrane that act as bearings; in addition, it is believed the lower two (MS and C rings) generate the torque on the shaft that rotates the hook. It is striking how closely in morphology the motor resembles an engine in a macroscopic machine.

The basic power for the motor comes from positive hydrogen ions (protons) driven across the membrane by a combination of a concentration gradient and electrostatic force. The cells own metabolism sets up the concentration gradient, and this forms a speed control so that the rpm of the motor can be adjusted. The detailed torque generating mechanism is complex, but it is thought to arise from a difference in the number of protein subunits in the M and C rings [14]. The C ring has more subunits, and a separate stud attached to the membrane swaps sequentially between unpaired protein subunits pulling around the C ring relative to the M ring. Interestingly, it is believed that Brownian motion is an important part of the process.

The performance of the motor is impressive. Without an attached flagellum, it can rotate at up to 17,000 rpm and up to 1000 rpm under load generating thrust with the flagellum driving the bacterium body at speeds of up to 50 μm/sec. To get this into perspective, it is about 50 bacterial body lengths per second, which is equivalent to a car traveling at 700 km/hr (400 mph). At present it is not possible to replicate this motor synthetically or to use one out of its cellular environment, but entire bacteria have been used to move beads 5 times larger than themselves after attachment by electrostatic forces [15]. The bacteria, used rather like outboard motors, were able to propel the beads at speeds of ~15 μm/sec. In addition, they could be turned off by introducing copper ions into the solution and re-started by adding ethylenediaminetetraacetic acid.

7.2 ON-BOARD PROCESSING FOR NANOMACHINES

Nanomachines, whether autonomous nanobots or machines in the nanofactory, require some form of on-board processing to carry out their functions. This could vary from simple feedback mechanisms to complex decisions based on assessing a number of input parameters. Biological systems use chemical processing all the way from basic cellular functions to the processing that goes on in our brains. Technology, on the other hand, has focused almost entirely on electronic processing. At the present time there is no known way to equip a nanobot with a sufficiently small computer capable of carrying out even simple programs. Current microelectronics is just too bulky; and although it is feasible, using nanotechnology, to build circuit elements like transistors just a few nanometers across (see Chapter 5), there is no known way to interconnect them and build a circuit at this scale.

Biological computers are not sufficiently well understood to replicate in devices, so at the moment there is no clear solution for the nanobot brain problem. In this period of looking for solutions, there is, however, a third option, and that is a mechanical computer. In the mid-19th century the inventor Charles Babbage in the United Kingdom designed mechanical computing machines that reached a very high level of sophistication including a fully programmable computer (the *analytical engine*). Although he never finished his analytical engine, it has since been reconstructed at the Science Museum, London [16]. This is a large machine weighing several tons, but it has been suggested since the mid-1980s by Eric Drexler that it may be possible to replicate machines like this with nanoscale components. Efforts to do this [17] are at present focused on micro–electrical–mechanical systems (MEMS) technology that fabricates micron-sized mechanical components in Si using technology borrowed from the microelectronic industry. This will not produce any further miniaturization relative to current microelectronics, but micromechanical computers would have some advantages such as low power consumption and the ability to operate in extreme environments. It may be possible. however, to take the components down to molecular sizes, in which case the size of the computer would match the dimensions of the rest of the nanobot. This would run up against the "fat fingers," "sticky fingers," and Brownian motion problems, but it is possible that we could learn enough from nature to actively use the Brownian motion to help the processor work.

7.3   MEDICAL NANOBOTS

In vivo nanosystems for drug delivery and diagnostics were discussed in Chapter 6; and although these are entirely "passive" systems, they can reach a high level of sophistication as shown in Fig. 6.1. A fully equipped nanovector can carry (a) drugs and contrast enhancers for MRI imaging, (b) permeation enhancers that help it pass through blood vessel walls, (c) protection from the body's immune system, and (d) targeting so it locates the right type of cell. Medical nanobots take this functionality a stage further and introduce some autonomous diagnostic capability and the possibility of communicating their findings to outside the body.

The type of device envisaged is a machine that is smaller than a red blood cell (i.e., less than 5 μm in its longest dimension) and is equipped with a variety of detectors including chemical, pH, temperature, and flow rate sensors. At this size it will easily pass along even the smallest blood vessels (capillaries) in the body. It should be able to communicate any findings to outside the body and, in more sophisticated implementations, receive instructions. The main advantage of having a robot *in situ* is that it will be sensitive to local *in vivo* conditions and locate the exact source of problems. The inclusion of some method for locomotion would increase the diagnostic ability of a device because it could, for example, having detected a chemical, swim up the concentration gradient to locate the source. Having found the problem, it could release drugs at the problem site either as a result of a decision made by an on-board processor or as a result of a remote instruction. In this approach, minute amounts of a drug administered by nanorobots at exactly the right spot would have a therapeutic effect with minimal side effects. The nanovectors described in Chapter 6 also deliver drugs to precisely the right spot, but the main difference is that they work as magic bullets; that is, they find the target, dispense the drug, and are then discarded. There needs to be a pre-diagnosed problem for nanovectors to be used while a nanobot would be a powerful diagnosis aid for early detection and prevention.

The motion of such a tiny object through fluid is very different from that of a macroscopic object. It is entirely dominated by the viscosity of the liquid (see Advanced Reading Box 7.1); and if a propelling force is switched on, the nanobot will reach terminal velocity immediately; thus in this environment, force is associated with velocity rather than acceleration.

ADVANCED READING BOX 7.1—REYNOLDS NUMBER FOR NANO-SCALE OBJECTS IN FLUIDS

The motion of an object through a fluid is characterized by the dimensionless Reynolds number, Re (proposed in 1883 by the British engineer Osborne Reynolds), which is the ratio of the momentum per unit area of the fluid flowing at a velocity v to its viscosity, η. The momentum per unit area is $\rho v D$, where ρ is the density of the fluid and D is the diameter of the flow:

$$\text{Re} = \frac{\rho v D}{\eta}.$$

The units of η are $N \cdot s/m^2$, which is the same as momentum per unit area, so Re is dimensionless. The Reynolds number is the same whether the fluid is flowing with a velocity v or an object is moving through a stationary fluid with velocity v, though in this latter case, D is the size of the object. For Re $\gg 1$ the inertia of the object dominates, while for Re $\ll 1$ the viscosity of the fluid dominates.

It is interesting to compare the Reynolds numbers for a human and a bacterium swimming through water, whose viscosity is $8.9 \times 10^{-4} \ N \cdot s/m^2$. Typical values for a human are $v = 1$ m/s and $D = 0.5$ m, while for the bacterium they are $v = 5 \times 10^{-5}$ m/s and $D = 10^{-6}$ m. These give Reynolds numbers of 5.6×10^5 and 5.6×10^{-5}, respectively—10 orders of magnitude apart! Even if we substitute the viscosity of molasses (870 $N \cdot s/m^2$) for the human swimmer, we still only get Re ~ 0.5. So for the bacterium its environment appears to be considerably more viscous than molasses and its motion is completely dominated by the viscosity of the fluid. When it turns on its propulsion, it will reach terminal velocity instantly. The viscosity of blood is $\eta = 0.0027 \ N \cdot s/m^2$ and is higher than that of water, but the difference is not significant for the calculation.

The power requirements could be met from a variety of sources including low-level RF power supplied from the outside (similar to mobile phone radiation) or on-board sources such as some kind of chemical fuel. Communication with the outside world could be by radio telemetry as is already used for medical implants such as long-term heart monitors.

There are very few proving technologies for medical nanobots and most studies are theoretical [18]. In recent years, pieces of nanotechnology have emerged that have started to assemble some of the technical requirements for a medical nanobot, and some of these are shown in Fig. 7.7. The propulsion could be via an attached bacterium with a flagellum that could be turned on and off by releasing ethylenediaminetetraacetic acid or Cu ions, respectively, as already described in Section 7.1 [15]. The chemical sensors could be sections of

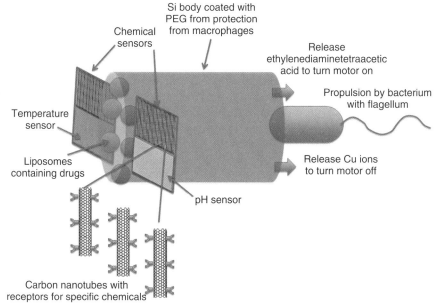

Fig. 7.7 Some components of a medical nanobot feasible with current technology. Schematic of a medical nanobot showing some components and functions that are feasible with current technology. The propulsion could utilize a bacterium with a flagellum that can be turned on and off by releasing ethylenediaminetetraacetic acid or Cu ions respectively [15]. The chemical sensors could be sections of conducting carbon nanotubes coated with receptors for specific chemicals (see Chapter 6, Section 6.2.4). Attachment of a single molecule of the chemical would produce a measurable change in the conductivity of the tube. Temperature and pH sensors are available using a variety of methods. The nanobot carries drugs in liposomes that would release on contact with a cell membrane (see Chapter 6, Section 6.1.2). A good deal is known about protecting the device from the bodies immune system from current research on nanovectors (see Chapter 6, Section 6.1.4).

conducting carbon nanotubes coated with receptors for specific chemicals (see Chapter 5, Section 5.4). Attachment of a single molecule of the chemical would produce a measurable change in the conductivity of the tube. The system would, however, saturate easily and there would need to be some mechanism (as yet not developed) to clear the receptors and reset the detector. Nanoscale temperature and pH sensors are available using a variety of current methods. The drugs could be carried in liposomes and, on arrival at a diseased cell, would be released simply by liposome contact with the cell membrane as described in Chapter 6, Section 6.1.2, though how the liposomes would be deployed is not known. A good deal is known already about protecting the device from the body's immune system from current research on nanovectors (see Chapter 6, Section 6.1.4). The major technologies missing are the processing system, the power to run the device, and communication with the outside world (or with other nanobots).

7.4 MOLECULAR ASSEMBLY

The only man-made atomic structures in which atoms have been deliberately placed at predetermined positions are like the ones shown in Chapter 4, Section 4.4.2, in which an STM tip has been used to painstakingly move atoms from one place to another on a surface. This is not a viable technology for molecular manufacturing, which requires some sort of nano-assembler working at the nanoscale. Again, to learn anything about molecular assembly, we need to turn to living organisms in which assembly at the molecular level is continuously happening within cells.

There are two types of molecular assemblers working in cells: (a) proteases, which manufacture DNA from nucleobases (see Chapter 6, Section 6.1.5), and (b) ribosomes, which manufacture proteins from a set of amino acid building blocks. The ribosome has the more complicated job because there are about 20 amino acid building blocks out of which to produce proteins as opposed to 4 nucleobases for DNA. Figure 7.8 shows the steps involved in a ribosome building a new protein according to instructions laid out along a messenger ribonucleic acid (mRNA) molecule. The mRNA is similar to DNA in that a set of four bases, labeled "A" (adenine), "G" (guanine), "C" (cytosine), and "U" (uracil), are strung along a phosphate backbone. In the case of mRNA the U base replaces the T (thymine) base found in DNA. The bases are arranged in groups of 3 (e.g., AUG) where each group is known as a *codon* and is a label for a specific amino acid. A section of mRNA with a sequence of 7 codons is shown in Fig. 7.8a.

The ribosome structure, shown schematically in Fig. 7.8c, consists of two subunits, labeled large and small and the mRNA feeds through it like a tape of instructions. The amino acids are brought to it by transfer RNA (tRNA); and if it has a codon that is complementary to an exposed codon on the mRNA, it is allowed to dock. The ribosome provides three docking positions denoted A (aminoacyl), P (peptidyl), and E (exit) along the mRNA tape for tRNA. The ribosome in Fig. 7.8c has started assembling a chain of amino acids (known as a

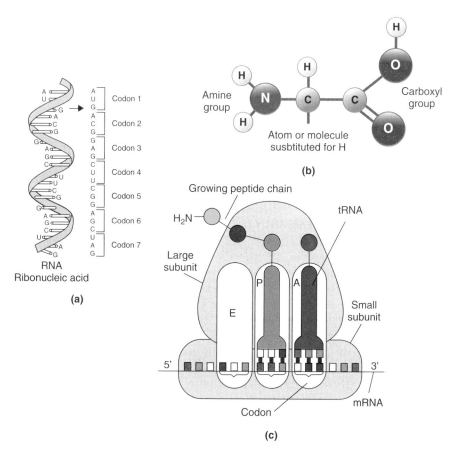

(a)

(b)

(c)

Fig. 7.8 Protein assembly by ribosomes. (a) A section of RNA in which a set of four bases, labeled 'A' (adenine), 'G' (guanine), 'C' (cytosine) and 'U' (uracil) are strung along a phosphate backbone. The bases are arranged in groups of 3 where each group is known as a *codon* and is a label for a specific amino acid. Reproduced from Wikipedia web page: http://en.wikipedia.org/wiki/Genetic_code (b) The generic structure of an amino acid. Reproduced from Wikipedia web page: http://en.wikipedia.org/wiki/Amino_acid (c) Schematic of a ribosome showing the 3 docking positions for transfer RNA (tRNA). The ribosome has started making a chain of amino acids represented by the colored circles and a tRNA carrying the next one (blue circle) as coded by the mRNA has just docked. Its amino acid will be added to the chain after which the mRNA and tRNA will translocate by one codon so that the tRNA at P moves to E (exit) the one at A moves to P and the next codon is exposed at dock A and waits for a tRNA carrying the correct amino acid. Reproduced with permission from Prof. J. Frank, Columbia University.

peptide chain), shown as colored circles. The tRNA docked at P has just provided the yellow amino acid; and according to the instructions on the mRNA, the next required building block is an amino acid corresponding to the blue circle. In this case the codon on the bottom of purple tRNA molecule is complementary and it has been allowed to dock. The amine group (Fig. 7.8c) of the amino acid attached to the tRNA docked at A reacts with the carboxyl group of the amino acid attached to the tRNA at P, and the peptide chain grows by one amino acid unit. After the reaction the mRNA translocates one codon so that the tRNA at P moves to the exit position, the A site tRNA moves to the P position, and the dock at A has the next codon exposed and waits for a tRNA with the correct amino acid attached to dock and repeat the process. On the mRNA there is a "finish" codon to signal that the peptide chain is complete and initiates a process to release it into the cell. A superb video of the ribosome producing a polypeptide chain is available on the Web [19].

This is a truly impressive assembler, and the entire ribosome packs into a roughly spherical shape 20 nm in diameter. It has evolved, however, to do a very specific job—that is, assemble a chain of amino acid building blocks to produce proteins. The construction of the equivalent of a ribosome that does this job for, say, the atoms of the 92 elements is totally beyond any foreseeable technology. It is possible, however, to "reprogram" the ribosome to produce new protein structures as demonstrated this year by George Church and Michael Jewett at the Harvard Medical School. They extracted natural ribosomes from *Escherichia coli* bacteria, broke them down into their constituent parts, removed the key ribosomal RNA, and then rebuilt the ribosome using RNA synthesized from molecular building blocks. Their new ribosome was able to manufacture a mirror image of a naturally produced protein. The short-term benefit will be industrial because natural ribosomes are already used to produce proteins so that modified versions will extend the range that can be manufactured. In the longer term the ability to produce "designed" protein structures will enable a whole host of nanoscale devices such as "boxes" that carry a molecular cargo and release it on demand with a chemical signal. This would be vital component of the nanofactory and could solve the problem posed at the end of Section 7.1—that is, how to use a microtubule to carry a cargo and the release it at the correct stage.

Thus looking at the nanofactory as a hierarchy working at different scales, we can see technology gradually moving down the scales. Top-down nanotechnology can produce complex working machines down to the micron level. We can harness biology to produce, for example, the conveyor belts utilizing kinesin and microtubules described in Section 7.1 down to the 100-nm level. Artificial ribosome technology could, at some point in the future, manufacture working devices based on proteins down to the 10-nm level. The technology to go beyond this point is simply missing. We need to learn much more about the natural world before any feasible solution is likely to present itself.

In the final chapter we will examine how nanotechnology is helping to increase our understanding of the quantum vacuum—the most fundamental system in the universe, which at small scales generates forces that we don't fully understand.

It also looks at the possibility of harnessing these forces as a partial solution to the "fat fingers" and "sticky fingers" problem of radical nanotechnology.

REFERENCES

1. http://www.youtube.com/watch?v=zqyZ9bFl_qg

2. R. E. Smalley, Of chemistry, love, and nanobots, *Sci. Am*. **285** (2001), 76–77.

3. A. Yildiz, M. Tomishige, R. D. Vale, and P. R. Selvin, Kinesin walks hand-over-hand, *Science* **303** (2004), 676–678.

4. http://www.scripps.edu/cb/milligan/projects.html

5. H. Hess, J. Clemmens, J. Howard, and V. Vogel, Surface imaging by self-propelled nanoscale probes, *Nano Lett*. **2** (2002), 113–116.

6. Y. Hiratsuka, T. Tada, K. Oiwa, T. Kanayama, and T. Q. P. Uyeda, Controlling the direction of kinesin-driven microtubule movements along microlithographic tracks, *Biophys. J*. **81** (2001), 1555–1561.

7. M. G. L. van den Heuvel, C. T. Butcher, S. G. Lemay, S. Diez, and C. Dekker, Electrical docking of microtubules for kinesin-driven motility in nanostructures, *Nano Lett*. **5** (2005), 235–241.

8. B. Yurke, A. J. Turberfield, A. P. Mills, Jr., F. C. Simmel, and J. L. Neumann, A DNA-fuelled molecular machine made of DNA, *Nature* **406** (2000), 605–607.

9. J. Bath and A. J. Turberfield, DNA nanomachines, *Nature Nanotechnol*. **2** (2007), 275–284.

10. T. Omabegho, R. Sha, and N. C. Seeman, A bipedal DNA Brownian motor with coordinated legs, *Science* **324** (2009), 67–71.

11. S. J. Green, J. Bath, and A. J. Turberfield, Coordinated chemomechanical cycles: A mechanism for autonomous molecular motion, *Phys. Rev. Lett*. **101** (2008), 238101.

12. J. Bath, S. J. Green, and A. J. Turberfield, A free-running DNA motor powered by a nicking enzyme, *Angew. Chem. Int. Ed*. **44** (2005), 4358–4361.

13. J.-S. Shin and N. A. Pierce, A synthetic DNA walker for molecular transport, *J. Am. Chem. Soc*. **126** (2004), 10834–10835.

14. D. R. Thomas, D. G. Morgan, and D. J. Derosier, Rotational symmetry of the C ring and a mechanism for the flagellar rotary motor, *Proc. Natl. Acad. Sci. USA* **96** (1999), 10134–10139.

15. B. Behkam and M. Sitti, Bacterial flagella-based propulsion and on/off motion control of microscale objects, *Appl. Phys. Lett*. **90** (2007), 023902.

16. See the Science Museum web page: http://www.sciencemuseum.org.uk/objects/computing_and_data_processing/1878-3.aspx

17. See, for example, the University of Wisconsin press release:http://www.news.wisc.edu/13981

18. A. Cavalcanti, B. Shirinzadeh, R. A Freitas, Jr., and T. Hogg, Nanorobot architecture for medical target identification, *Nanotechnology* **19** (2008), 015103.

19. http://www.youtube.com/watch?v=Jm18CFBWcDs

CHAPTER 8

Prodding the Cosmic Fabric

8.1 ZERO-POINT ENERGY OF SPACE

In Chapter 4 the tools of nanotechnology were described and their ability to build tiny structures varying in size from single atoms up to the edge of the nanoworld (100 nm) was demonstrated. It has been discovered recently that these same tools can address a much more fundamental question that takes us back to the discussion in Chapter 1, regarding the nature of "the void." Remember that the concept of the atom as a building block of matter, proposed by Leucippus and Democritus 2500 years ago, requires as part of the package, a "void" in which the atoms move. In fact it was probably the philosophical problems associated with having a void that killed off the ancient atomic theory, which is a tragic lost opportunity, because the modern version did not reemerge until the work of John Dalton 2300 years later.

Although atoms are now a familiar part of the scientific landscape, the true nature of the void remains somewhat mysterious. In classical physics the void is assumed to be a simple absence of all matter and all energy—pure space that has been emptied of everything detectable. This idea of pure space, as an undetectable container of all things we *can* detect did not just cause the ancient Greeks problems. More recently, Newton struggled with the same concept and in the end simply defined it:

> I do not define time, space, place, and motion, as they are well known to all. Absolute space by its own nature, without reference to anything external, always remains similar and unmovable.

In the same way, Newton also thought in terms of an absolute time. Later statements showed that he was not altogether happy with these definitions, which were more a matter of pragmatism so that he could get on with developing the rest of his scientific framework. Einstein showed that empty space is not just the three dimensions of space but has to incorporate the time dimension to

Introduction to Nanoscience and Nanotechnology, by Chris Binns
Copyright © 2010 John Wiley & Sons, Inc.

produce empty "space–time," which itself "curves" to generate gravity. The idea of an empty container of matter and energy, whether it is composed of three or four dimensions, is becoming increasingly hard to maintain in the face of the newly discovered "dark matter" and "dark energy" that appear to be invisibly present in the empty container even when everything detectable has been removed.

A more accurate picture of the void that has emerged from quantum mechanics—or more specifically, quantum field theory—is that rather than being an empty container, space or space–time is a "something" (precisely what, remains a mystery) that itself generates all the particles and energy that we can detect as a result of being stimulated in the right way. Energy manifests itself as particles; thus in the following discussion, "particle" means both matter and energy. According to the modern view, the "something" that is the void is a set of "fields," each of which has an intrinsic, or "zero point," energy and is able to generate its own specific type of particle as an "excitation." Thus the electromagnetic field makes photons, the electron or "Dirac" field makes electrons, and so on. A simple analogy is to imagine a perfectly still ocean with no waves. If it is stimulated with a very simple excitation, say a small pebble, it produces a wave, which in the quantum view is a single particle. Suppose we somehow had produced a very complicated stimulation that produced huge numbers of particles that interacted and combined in such a way as to produce a conscious intelligence. This conscious intelligence could build particle detectors that would detect simple waves and set up a formalism to describe them. To this intelligence, there would be particles, or simple waves, and an absence of all detectable waves, which it would call the void. For another being that lived in more dimensions and could see the whole picture, however, this absence of waves would not be a void, it would be a vast ocean—just calm. In the same way, quantum field theory tells us that an absence of particles and energy (or the void) is simply the condition when all the fields are quiet and unstimulated. In this condition, all the fields still contain an intrinsic energy, the "zero-point" energy, that we have no way of detecting directly.

In this way of looking at things, we can revisit the idea of the atom and the void. In the period covering the time of Democritus through to the development of quantum field theory in the 1930s and 1940s, the atom was a particle (as we now know, itself composed of smaller particles) moving in a void. For the last few decades the scientific picture is that a few excitations in the fields that compose the void generate the electrons, protons, and neutrons that make up the atom, which interact with each other by exchanging force particles that are themselves generated by the void. This clump of excitations is the atom, and it is not something separate that moves around in a void. It is itself part of the void. Figure 8.1 provides a simple illustration of the different points of view.

Let us now focus the discussion on the zero-point energy of the fields that compose the void; to simplify matters, consider just one field, the electromagnetic field, that makes photons. These are the fundamental particles of light and one can consider a beam of light as being a beam of photons, each one of which is

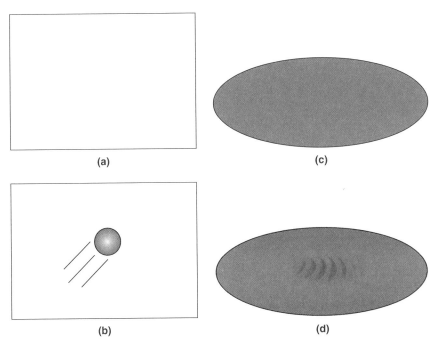

Fig. 8.1   Classical and quantum view of the void. (a) In classical physics, the void is an empty, but nevertheless real, container that provides an absolute reference for motion. (b) A particle is a separate object that can move through the empty container. (c) According to quantum field theory the void is a set of fields, with a huge zero-point energy density that are capable of vibrating, with an elementary vibration corresponding to a particle. For example, the electromagnetic field can manufacture photons. The perfect vacuum is the condition in which the field is not vibrating. (d) A single particle is a single elementary vibration (excitation) in the field.

an excitation in the underlying electromagnetic field. So what is the nature of the raw unexcited field with no photons? Again a simple analogy may help here. You can consider the electromagnetic field as being a quantum piano, where each string can vibrate to produce photons (Fig. 8.2). Since the field can vibrate at any frequency, the quantum piano has an infinite number of strings infinitesimally close together. The thing that makes it a quantum piano is that the strings can vibrate only at fixed amplitudes ascending in a sequence. If we then pluck just one string so it vibrates at its smallest allowed amplitude, this corresponds to a single photon at the string frequency. If we pluck it slightly harder so it vibrates at its next higher allowed amplitude, this corresponds to two photons at that frequency. If we hit it with a wide hammer that stimulates many strings with varying amplitude, we create a multitude of photons at a range of frequencies. This is what we are doing when we switch on a torch. The chemical energy in the battery heats a wire, and the vibrating atoms in the wire stimulate the electromagnetic field to create a large number of photons.

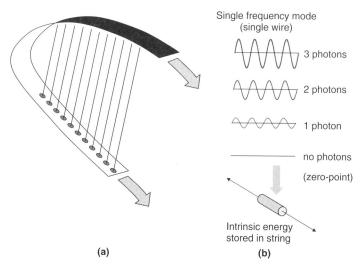

Fig. 8.2   Quantum piano. (a) The electromagnetic field can be represented by a 'quantum piano' where, each frequency that the field can vibrate at is represented by one string. There would have to be an infinite number of strings infinitesimally close together since the field can vibrate at any frequency. (b) A single wire represents a single vibration frequency or 'mode' of the field and since it is a quantum field the vibration amplitude can only take certain values. Thus the lowest possible vibration amplitude corresponds to a single quantum, which manifests as a particle, that is, a photon in the case of the electromagnetic field. The second lowest allowed vibration amplitude is two photons in that mode etc. A string not vibrating at all generates no photons but it still has an intrinsic tensile energy in the string, which gives it the capability of vibrating. This can be considered as the zero-point energy of the string or field mode. Thus a completely quiet piano, generating no particles constitutes the void but it still contains an infinite zero-point energy.

Suppose we now quieten everything down so that none of the strings are vibrating. According to quantum field theory, each string still has an intrinsic or zero-point energy; and in the case of our piano, this can be considered to be the tensile energy stored in the string that gives it the capability to vibrate. Each string has a fixed zero-point energy; and since there are an infinite number of strings (or modes), the zero-point energy of the whole piano is infinite. It is the same with the electromagnetic field. Each mode, or frequency at which the field can vibrate, has a fixed zero-point energy, which can be considered to be the energy that gives the field the capability to vibrate. So with an infinite number of modes the field has an infinite zero-point energy.

Whenever an infinity is encountered in describing a physical system, it is an indication that something is wrong with our conception of the system or something is missing. This isn't simply a result of our limited brains being unable to imagine the concept; it is also due to the fact that infinity introduces an arbitrariness that removes our ability to describe a system. The possible missing element in this case is the so-called Planck scale, which sets an ultimate limit on the smallest length that has meaning (see Advanced Reading Box 8.1), which is about 10^{-35} m.

ADVANCED READING BOX 8.1—PLANCK SCALE

The Planck scale is derived from the universal constants, c, the speed of light, h, the Planck constant, and G, the gravitatational constant [1]. Of all the constants used in science, these are thought to be the most fundamental since every other constant can be defined in terms of them. The combination of these three—defined by $c^x G^y h^z$, where x, y, and z are rational numbers—can take on the units of length, time, mass, and so on, by a suitable choice of x, y and z. For example, a "Planck length," l_{pl}, can be defined by

$$l_{pl} = (hG/c^3)^{1/2} = 10^{-35} \text{m}.$$

This is believed to be the smallest distance that can be defined within a physical system and may represent a fundamental "graininess" of space.

The Planck length thus also defines the shortest wavelength (highest frequency) photon that can exist (that is, our quantum piano has a shortest string) and this brings the zero-point energy to a finite but staggeringly large value of 10^{115} J/m^3 (see Advanced Reading Box 8.2). Just to get this in perspective, it suggests that the total energy output throughout their entire life of all the stars in all the galaxies in our observable universe could be obtained from a region of empty space considerably smaller than an atom. It is not infinity but still way beyond imagination. This huge number, derived from one pillar of modern science, is also in massive contradiction to the average energy density of empty space of 10^{-9} J/m^3 derived from the other pillar of modern science—that is, general relativity. This discrepancy by more than 120 orders of magnitude between the two great formalisms remains something of an embarrassment. Modern unification theories, such as string theory [2], are attempting to solve this discrepancy and unite the two theories.

ADVANCED READING BOX 8.2—ZERO-POINT ENERGY DENSITY OF THE ELECTROMAGNETIC FIELD

The zero-point energy of the electromagnetic field, E_0, arises from half a quantum in every possible field mode, that is,

$$E_0 = \sum_{kl} \frac{1}{2} \hbar \omega_\mathbf{k},$$

where **k** is the wavevector of the mode and l, represents the two possible polarizations. If we apply cyclic boundary conditions over a cube (and then let the size of the cube go to infinity), the sum over discrete k vectors transforms into the integral:

$$\sum_{kl} \frac{1}{2} \hbar \omega_\mathbf{k} = \sum_l \frac{V}{8\pi^3} \int \frac{1}{2} \hbar \omega_\mathbf{k} d^3 k.$$

Thus the energy density, e_0, taking into account that there are two photon modes (l values) per k value is

$$e_0 = \frac{1}{8\pi^3} \int \hbar \omega_k d^3 k.$$

Using $\int d^3 k = \int 4\pi k^2 dk$ and $k = \omega_k/c$ and $dk = d\omega_k/c$, this becomes

$$e_0 = \frac{4\pi \hbar}{8\pi^3 c^3} \int \omega^3 d\omega.$$

Thus the spectral variation, $\rho_0(\text{w})$, of the zero-point energy density is given by

$$\rho_0(\omega) = \frac{\hbar \omega^3}{2\pi^2 c^3}.$$

The zero-point energy density up to a maximum cutoff frequency, ω_{max}, say, is thus

$$\int_0^{\omega_{max}} \rho_0(\omega) d\omega = \frac{\hbar \omega_{max}^4}{8\pi^2 c^3}$$

The Planck length ($\sim 10^{-35}$ m) probably represents a natural limiting wavelength for photon modes, and this corresponds to $\omega_{max} \sim 10^{44}$ rad/s. Substituting this into the equation above gives a zero-point energy density of 10^{115} J/m^3.

8.2   THE CASIMIR FORCE

This esoteric discussion may seem far removed from nanotechnology, but it turns out that there are subtle measurable consequences of the zero-point energy that require the tools of nanotechnology to detect them; and in an ironic twist, these same subtle consequences could turn out to be the answer to the problems of nanomachines discussed in Chapter 7. In 1948, a Dutch physicist, Hendrik Casimir, pointed out that placing two perfect mirrors facing each other in empty space would disturb the zero-point electromagnetic field. This is because in such a region, only certain wavelengths (those that fit exactly into the space between the mirrors) are allowed. If only certain wavelengths of real photons are allowed, then something must have happened to the zero-point field that forbids it to generate other wavelengths of photon. It is as if something had removed strings from the quantum piano. The zero-point energy density must therefore be lower in the region between the mirrors than outside, and this difference in energy will depend on the distance between the mirrors. An energy that depends on distance equals a force, and so Casimir's simple prediction was that two mirrors in empty space would attract each other [3].

This appears at first hand to be an easy thing to test, especially because the force can be quite large at very small distances; for example, for a 10-nm separation of the mirrors, the pressure on them is the same size as atmospheric pressure (10^5 N/m^2). The force, however, drops very rapidly (as the inverse fourth power of separation for perfect mirrors) and becomes very hard to measure at separations above 1 μm. Until recently, it has been technically impossible to set up an experiment to accurately measure the force between two sufficiently flat and sufficiently smooth parallel surfaces that can approach to submicron distances. An early attempt by Marcus Sparnaay to measure the Casimir force in 1958 using a spring balance [4] succeeded in confirming its existence but was not able to measure it with sufficient accuracy to compare its magnitude with theory. Casimir's theory made the assumption that the mirrors were perfect—that is, were totally reflecting at all wavelengths. The theory was extended in 1956 by Lifshitz [5] to include real mirrors made out of real materials. In 1997, Steve Lamoreaux, at the University of Washington, Seattle, measured the Casimir force between a metal sphere and a metal plate using a very sensitive torsion balance [6]. The sphere–plate geometry (see Fig. 8.3) produces a smaller Casimir force at a given separation, but it can still be rigorously calculated and it gets rid of one of the big experimental problems with flat plates—that is, maintaining perfect parallelism down to tiny separations. Lamoreaux's measurement agreed with theory to a precision of better than 10% and the strange Casimir force arising from nothing, but empty space became a rigorously established experimental reality.

Enter the tools of nanotechnology described in Chapter 4. A device for measuring small forces at small separations *par excellence* is the atomic

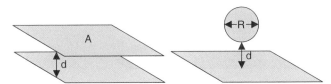

Fig. 8.3 Plate-plate and sphere-plate casimir force measurements. (a) For perfect parallel reflectors the Casimir force depends only on the area of the plates, A, the distance between them, d, and the universal constants, c (the speed of light) and h (Planck's constant). (b) In the sphere-plate geometry the Casimir force depends only on the diameter of the sphere, the gap between the sphere and the plate and the same universal constants. The magnitude of the force, for a given value of d is less in the sphere-plate configuration, however it removes one of the experimental problems, that, is maintaining perfect parallelism of two flat plates at sub-micron separations.

force microscope (AFM; see Section 4.4.4). Its utilization in Casimir force measurements was pioneered by Umar Mohideen at the University of California, Riverside [7] (see Fig. 8.4); and since 1998 a raft of measurements between spheres and plates using AFMs has been reported and the agreement between theory and experiment has steadily improved though the actual figure for the accuracy is still generating a discussion in the literature. The quest for

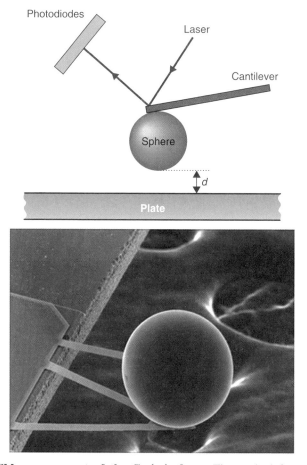

Fig. 8.4 AFM measurement of the Casimir force. The method for measuring the Casimir force with an AFM is shown schematically in the top figure. The AFM cantilever has a gold-coated sphere glued onto the end instead of having a sharp tip as normally used to scan a surface topography (see Chapter 4, Section 4.4.4). The deflection of the cantilever is determined in the usual way, that is, a laser beam is reflected from the back of a cantilever and the deflection of the reflected beam measured by a position sensitive detector. The bottom picture shows the 200 μm diameter gold-coated polystyrene sphere and cantilever used in the Mohideen experiment [7]. Reproduced with permission from the Institute of Physics pages *physicsweb* (http://physicsweb.org/articles/world/15/9/6).

greater precision in the measurements is important because it can set limits on parameters in new unification theories attempting to combine general relativity and quantum mechanics and develop a theory of quantum gravity. Thus the AFM, a tool of nanotechnology, has become a probe of fundamental aspects of the quantum vacuum.

8.3 THE CASIMIR FORCE IN NANOMACHINES

From the opposite point of view, the Casimir force has become of great practical significance in micro- and nanoscale mechanical devices. As the size of these devices has decreased, they have become full of boundaries with submicron gaps where the Casimir force becomes dominant. In fact it is a significant problem because while one can take measures to prevent things like capillary and electrostatic forces, there is nothing that can be done to prevent the Casimir force as it arises from the fundamental properties of the vacuum. It thus generates a fundamental and ever-present stickiness of components in micromachines and nanomachines.

More recently, research on Casimir forces has started to investigate the possibility of turning the problem on its head and utilizing the Casimir force as a useful method to transmit force without physical contact. The first real demonstration of the utilisation of the Casimir force to modify the motion in a micromechanical system was performed in 2001 by Frederico Capasso and his team at Bell Laboratories, New Jersey [8]. They built a standard micromechanical device consisting of a flat silicon plate with dimensions of a few hundred microns, which is suspended by a torsion wire above a surface (see Fig. 8.5). By applying an AC voltage to the pads underneath the plate, it can be made to oscillate seesaw fashion at a frequency of a few kilohertz. Then, using an AFM-type manipulator, they lowered a 100-μm gold-coated sphere toward one side of the oscillating plate approaching it to within a few hundred nanometers. The Casimir force between the sphere and the plate causes a shift in the frequency of the oscillator. They found that the amplitude and frequency shift of the oscillator were measurable with only a few nanometers change in the height of the sphere.

Another important step toward utilizing the Casimir force was taken in 2002 when the Mohideen group demonstrated a lateral force between a corrugated surface and a similar corrugation imprinted onto a gold-coated sphere [9] (see Fig. 8.6). Thus when the sphere is moved parallel to the corrugation, a lateral force that tends to drag the corrugated surface in the same direction is generated through the vacuum.

Could this be the answer to the stickiness problem in nanomachines? The basic idea for a simple rack and pinion is shown in Fig. 8.7. A rack is moved laterally at a distance of a few tens to hundreds of nanmometers away from a pinion, and a force is transmitted through the vacuum to rotate the pinion. It could be as simple as finding the right combination of materials and the right shape of corrugations to make this kind of machine a practical reality. In fact the lateral Casimir force may

Fig. 8.5 Using the Casimir force in micromachines. Demonstration of utilization of the Casimir force to modify the frequency of a micro-mechanical oscillator by using AFM technology to bring a gold-coated sphere to within a few hundred nanometers of one side of a torsion oscillator [8]. Reproduced with permission from Prof. Federico Capasso, Harvard.

be of far more utility than a passive force transmission mechanism. The theory of the Casimir force in cavities is highly developed only for very simple geometries such as the plate–plate and sphere–plate. For a complex system such as the one shown in Fig. 8.7, it is still under development. A recent paper [10], however, indicates that the strength of the force transmission and even the direction of the force depend on the speed at which the rack is moved. Thus if it is moved backwards and forwards at different speeds in each direction, it may be possible to provide a continuous unidirectional torque on the pinion.

   This is particularly useful action mechanically because it is a method to transfer oscillatory motion into a unidirectional linear motion. This is what a car transmission system does when considered all the way from the oscillating pistons through to the linear motion of it driving up the street. An alternative method of achieving this has been suggested by another recent paper [11], and that is to define a system with an asymmetric set of teeth as shown in Fig. 8.8. Then oscillating the two surfaces in the normal direction will provide a unidirectional

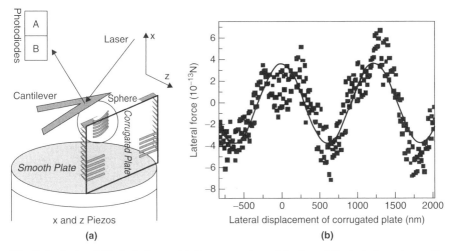

Fig. 8.6   Lateral Casimir force. (a) Demonstration of the lateral Casimir force transmitted through vacuum between a corrugated surface and a sphere with the same corrugation imprinted onto it. (b) Pulling it past the surface generates a lateral force that oscillates as the corrugations move past each other. Reprinted with permission from [9]. Copyright 2002 by the American Physical Society.

Fig. 8.7   Non-contact rack and pinion. The lateral Casimir force can be used as a method of transmitting force without contact as in this rack and pinion.

linear force, whose direction can be varied by altering the lateral displacement of the two sets of teeth. This is the so-called *Casimir ratchet* effect.

These are the early days in the research on utilizing the Casimir force for contactless transmission; but they hold the promise of eliminating the stickiness problem in micromechanical and nanomechanical components, allowing machines to continue to be scaled down toward molecular dimensions. Of course as gaps in a "contactless" transmission get smaller and approach the nanoscale, at some point it is reasonable to ask what "contact" actually means. That is, in the system shown in Fig. 8.7, everything is free to move whereas mechanical components in "contact" stick together. How close together do they have to be for

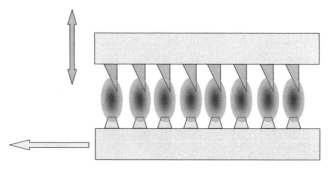

Fig. 8.8 Casimir ratchet. Using asymmetric teeth on one side, an oscillatory motion in a normal direction can produce a unidirectional motion of the other side.

everything to gum up? There is an answer to this question, though it is fairly hand waving. For two perfectly clean surfaces brought right into contact, they stick together due to a force known as the van der Waals force, which acts between any two atoms. In fact it is a subtle distinction between this and the Casimir force because they both spring from the same source—that is, the zero-point electromagnetic field. Their theoretical description is different, however, because generally the van Der Waals force is described in terms of surface charges on the two bodies whereas the Casimir force is described in terms of the zero-point electromagnetic field. The surface charges, however, are a result of fluctuations in the zero-point field; hence fundamentally the two forces arise from the same source. As two surfaces are brought together, the power law describing how the force varies with distance changes, and this is the crossover between the two regimes. Within the context of this chapter, this crossover can be thought of as the minimum distance between two components before they stick together. This distance is about 10 nm, so the machine shown in Fig. 8.7 could be scaled down until the teeth were about the size of large molecules.

In summary, it is clear that the tools of nanotechnology, such as scanning probe microscopes and micromachines, can provide important information about the fundamental nature of space, especially the zero-point electromagnetic field. An exciting aspect of this subject is that a better understanding of the force that arises from the zero-point field (i.e., the Casimir force) may enable its control to some extent. This would then feed back into improved micro- and nanomachines that utilize the force to achieve contactless transmission or exploit the Casimir ratchet effect.

REFERENCES

1. M. Planck, *Über Irreversible Strahlungsvorgänge. Fünfte Mitteilung*, Königlich Preussische Akademie der Wissenschaften (Berlin), Sitzungsberichte, 1899, pp. 440–480.

2. For a good description of string theory aimed at nonspecialists, see B. Greene, *The Fabric of the Cosmos*, Alfred A. Knopf Publishers, New York, 2004.

3. H. B. G. Casimir, On the attraction between two perfectly conducting plates, *Proc. R. Netherlands Acad. Arts Sci.* **51** (1948), 793–795.

4. M. J. Sparnaay, *Measurements of attractive forces between flat plates*, *Physica* **24** (1958), 751.

5. E. M. Lifshitz, *Sov. Phys. JETP* **2** (1956), 73.

6. S. K. Larmoreaux, Demonstration of the Casimir force in the 0.6 μm to 6 μm range, *Phys. Rev. Lett.* **78** (1997), 5–8.

7. U. Mohideen and A. Roy, Precision measurement of the Casimir force from 0.1 to 0.9 μm, *Phys. Rev. Lett.* **21** (1998), 4549.

8. H. B. Chan, V. A. Aksyuk, R. N. Kleiman, D. J. Bishop, and F. Capasso, Nonlinear micromechanical Casimir oscillator, *Phys. Rev. Lett.* **87** (2001), 211801.

9. F. Chen and U. Mohideen, Demonstration of the lateral Casimir force, *Phys. Rev. Lett.* **88** (2002), 101801.

10. A. Ashourvan, M. irFaez Miri and R. Golestanian, Non-contact rack and pinion powered by the lateral Casimir force, *Phys. Rev. Lett.* **98** (2007), 140801,

11. T. Emig, Casimir-force-driven ratchets, *Phys. Rev. Lett.* **98** (2007), 160801.

GLOSSARY

ADP	Adenosine diphosphate
ATP	Adenosine triphosphate
AFM	Atomic force microscope
DNA	Deoxyribonucleic acid
DPN	Dip-pen nanolithography
DTR	Data transfer rate
EBL	Electron beam lithography
FET	Field effect transistor
FIB	Focused ion beam
LED	Light emitting diode
MFM	Magnetic force microscope (or microscopy)
MRSA	Methicillin resistant *Staphylococcus aureus*
MWNT	Multi-walled nanotube
RNA	Ribonucleic acid (tRNA; transfer RNA; mRNA; messenger RNA)
SAM	Self-assembled monolayer
SAR	Specific absorption rate
SEM	Scanning electron microscope (or microscopy)
SET	Single-electron transistor
SPM	Scanning probe microscope (or microscopy)
SPR	Surface plasmon resonance
STEM	Scanning transmission electron microscope (or microscopy)
STM	Scanning tunnelling microscope (or microscopy)
SWNT	Single-walled nanotube
TEM	Transmission electron microscope (or microscopy)

For entries that appear in more than one page or section in the book, the page numbers in bold indicate the pages on which the main reference to the entry can be found.

Introduction to Nanoscience and Nanotechnology, by Chris Binns

Copyright © 2010 John Wiley & Sons, Inc.